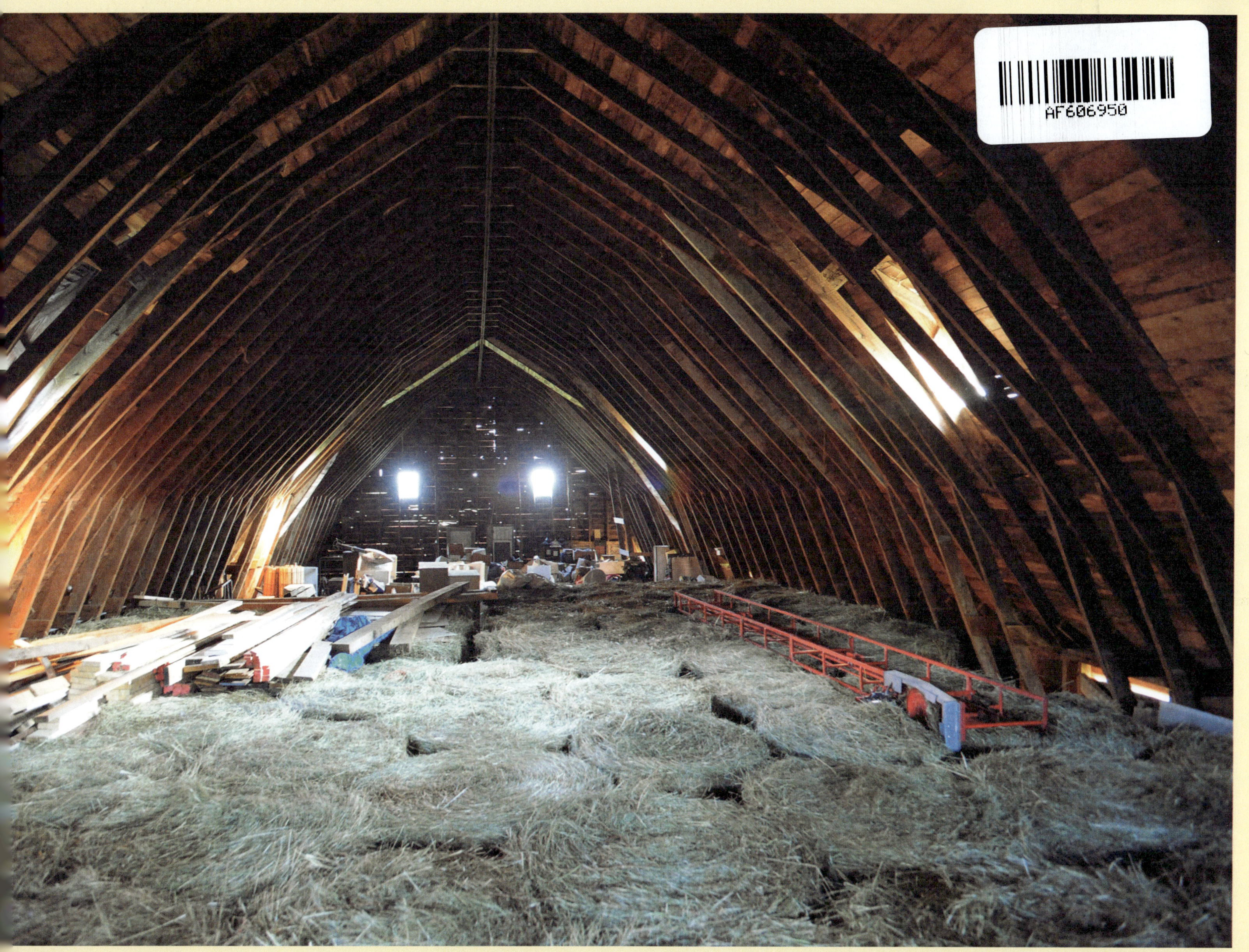

Barns of the Berkshires

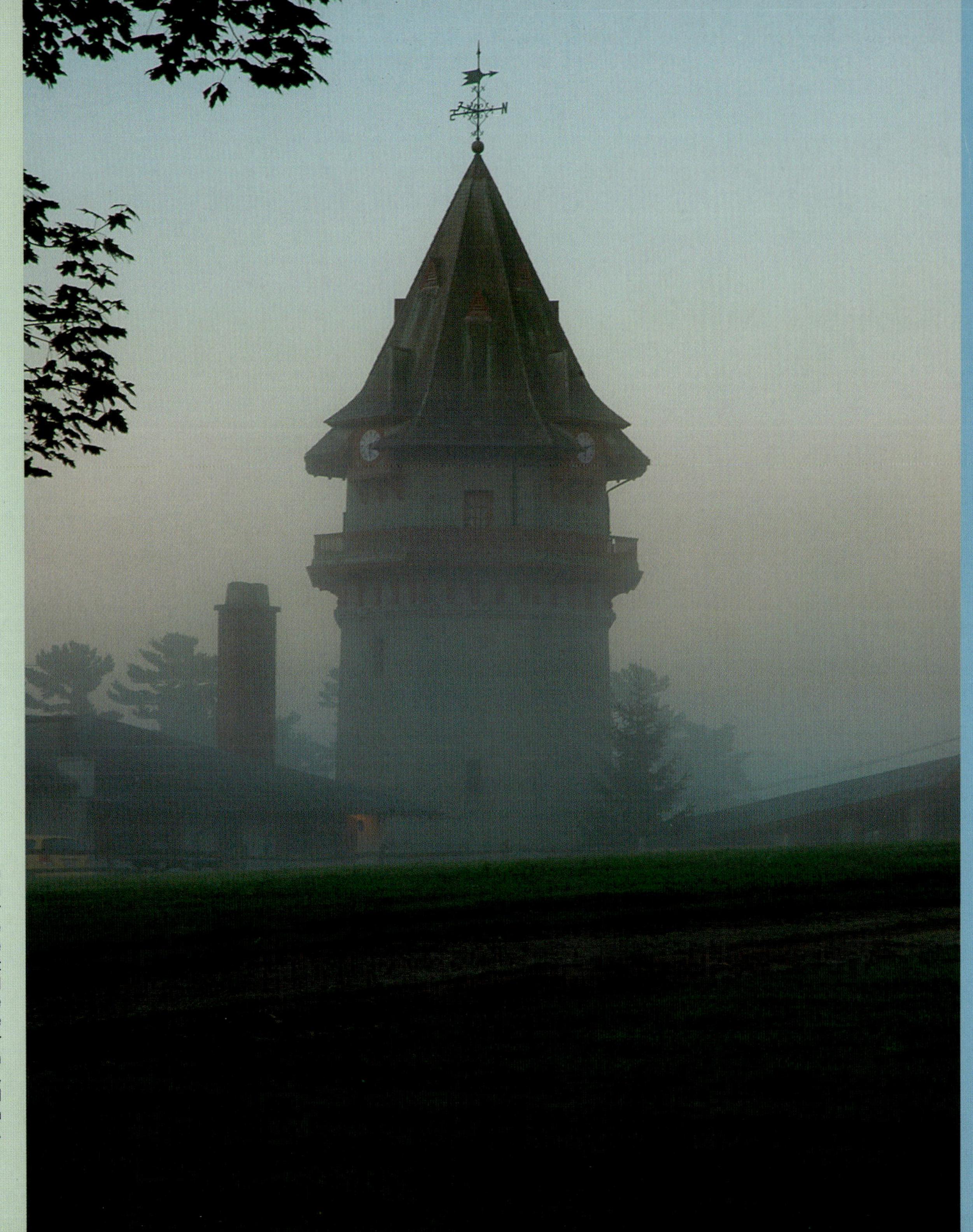

The dairy at High Lawn Farm has been operated by the same family for over 100 years and is the last remaining complete dairy farm—meaning that it produces, bottles, and delivers its own milk—in the Berkshires. Over the years the clock tower that stands over the barn complex, shown here in the mist at sunrise, has become a recognized icon of the area by both residents and visitors.

Barns of the Berkshires

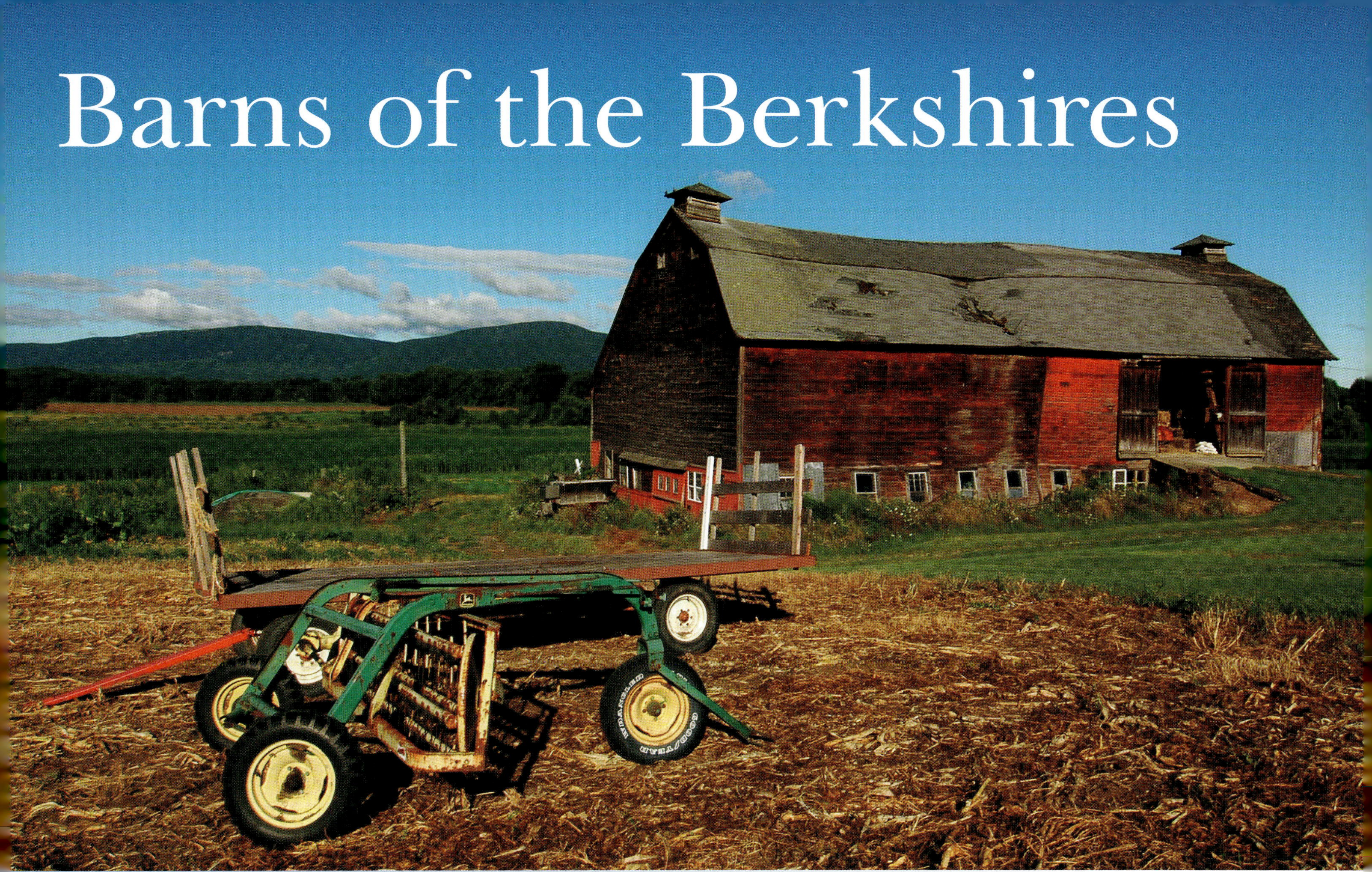

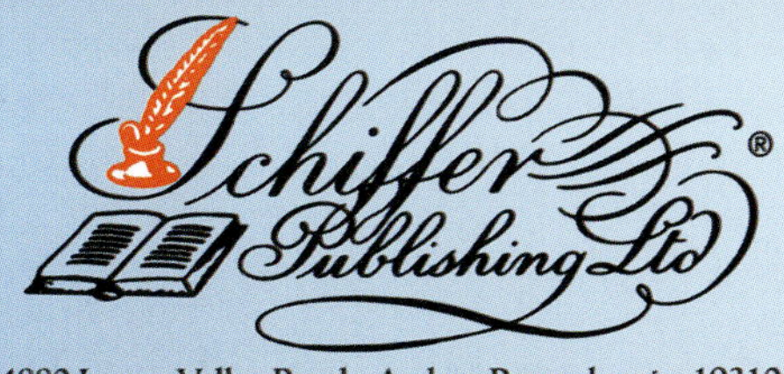

4880 Lower Valley Road, Atglen, Pennsylvania 19310

Stephen G. Donaldson

Dedication

This book is dedicated to the farmers of Berkshire County whose essential role in our daily lives, and the hard work it requires, goes largely unacknowledged, and to the barns that stand as noble monuments to the ingenuity, perseverance, and courage that helped establish a New World.

Cover Photo: These two famous barns are located on East Road in Adams on the Golden Acres Farm property of Blanche Demagall.

Title Page: This large gambrel roof barn at Dick and Carol Ann Boardman's farm once housed 75 head of cattle. Built c. 1894, it was added to in 1938. Like many late-18th century barns, it is not a post and beam structure.

Library of Congress Control Number: 2008941376

Designed by Mark David Bowyer
Type set in New Baskerville / Zurich BT

ISBN: 978-0-7643-3223-4
Printed in China

Schiffer Books are available at special discounts for bulk purchases for sales promotions or premiums. Special editions, including personalized covers, corporate imprints, and excerpts can be created in large quantities for special needs. For more information contact the publisher:

Published by Schiffer Publishing Ltd.
4880 Lower Valley Road
Atglen, PA 19310
Phone: (610) 593-1777; Fax: (610) 593-2002
E-mail: Info@schifferbooks.com

For the largest selection of fine reference books on this and related subjects, please visit our web site at **www.schifferbooks.com**
We are always looking for people to write books on new and related subjects. If you have an idea for a book please contact us at the above address.

This book may be purchased from the publisher.
Include $5.00 for shipping.
Please try your bookstore first.
You may write for a free catalog.

In Europe, Schiffer books are distributed by
Bushwood Books
6 Marksbury Ave.
Kew Gardens
Surrey TW9 4JF England
Phone: 44 (0) 20 8392-8585; Fax: 44 (0) 20 8392-9876
E-mail: info@bushwoodbooks.co.uk
Website: www.bushwoodbooks.co.uk
Free postage in the U.K., Europe; air mail at cost

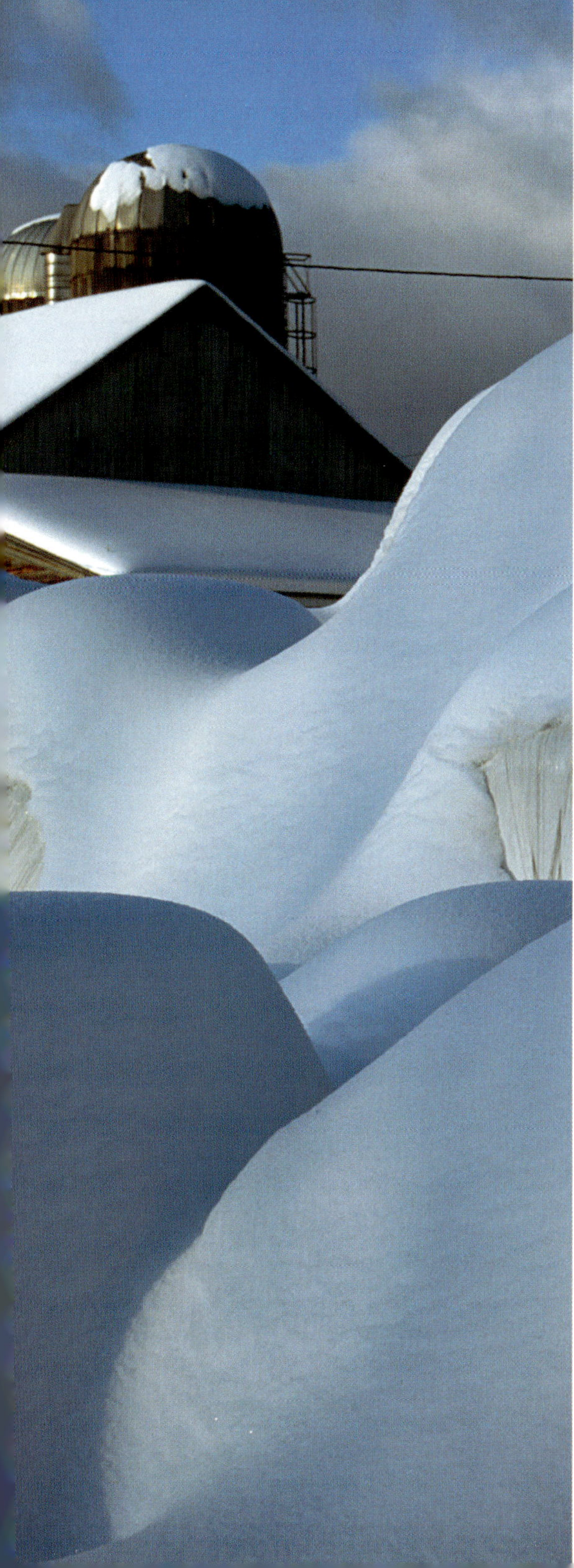

Contents

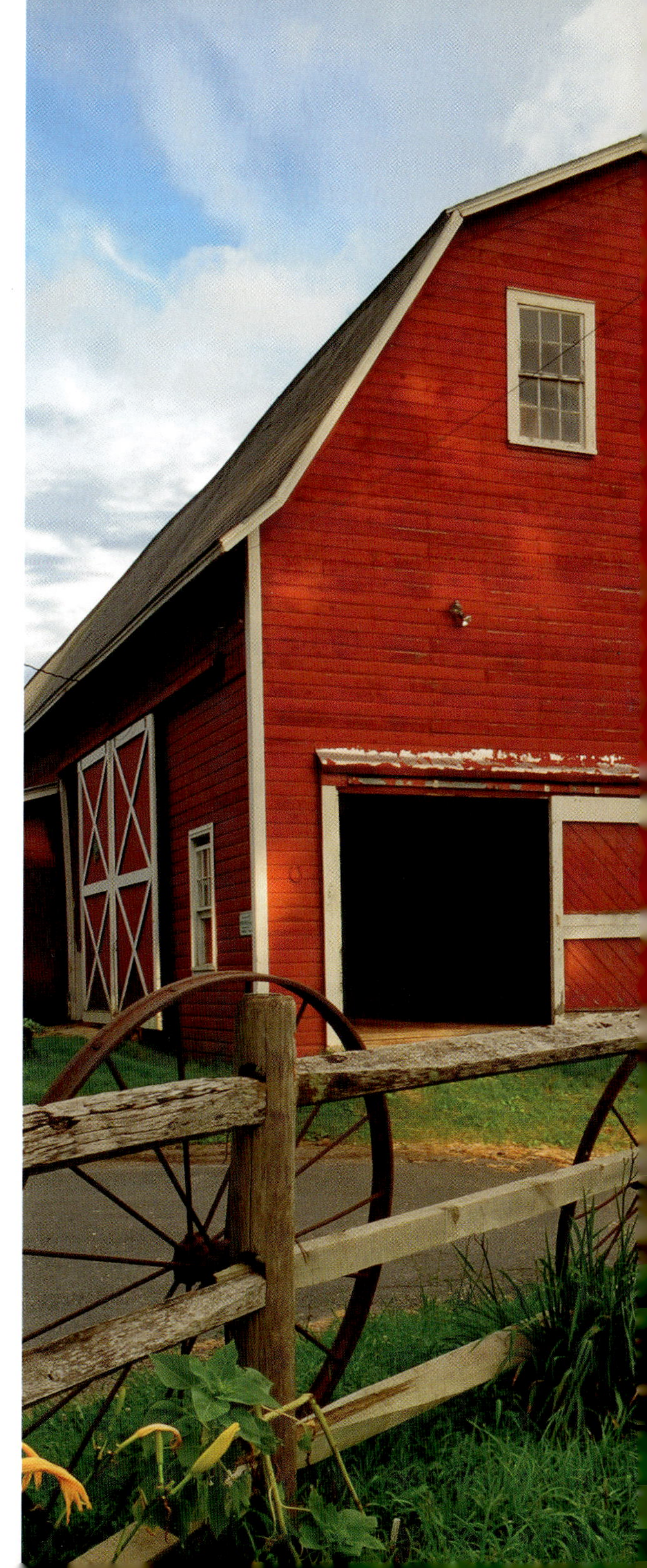

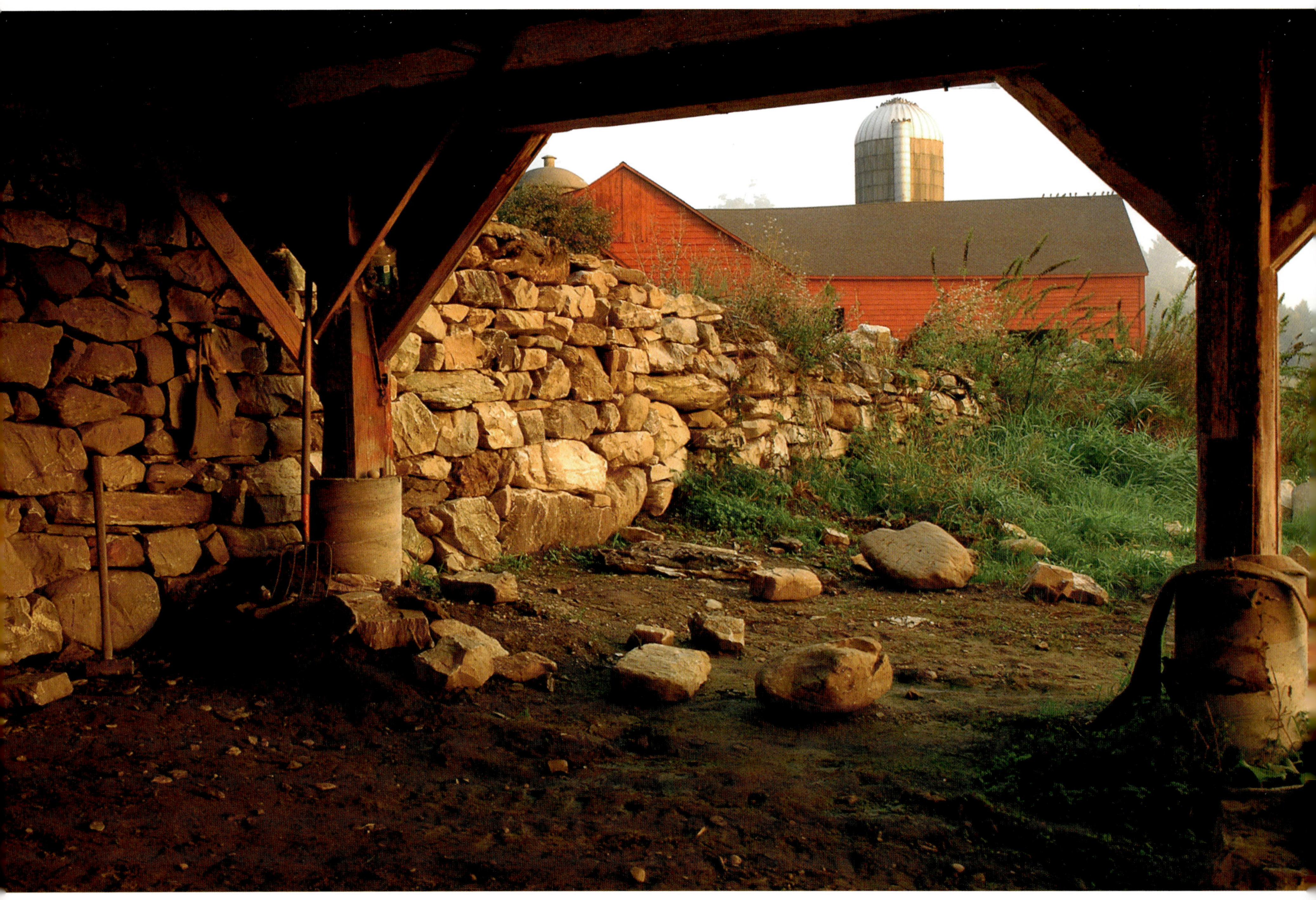

Acknowledgments

This book would not have been possible without the kindness, cooperation, understanding, patience, knowledge, memory, and expertise of countless people I have met in the course of compiling the material for it.

Most of all, I am grateful to all of the people who opened their hearts, homes, and barns to me so that I could photograph all of the magnificent buildings highlighted here. Some of the photographs did not make it into the book, but that does not negate my appreciation of the cooperation of their owners. I will list them in no particular order: Charles Proctor, Gary Shepard, Lila Berle, Dick and Carol Ann Boardman, Rachel Fletcher, Barbara Delmolino, Richard and Priscilla Burdsall, Beth Bartholomew, Richard and Pam Armstrong, Don and Alice Hale, Daniel Menaker, Ron Majdalany, Ed Clairmont, John Curtin, Peter Curtin, Ken Demers, Warren Wood, Marlene and Howard Tanner, Jack Sobon (whose depth of knowledge of the English barn, and willingness to share it, was indispensable), Dave Lanoue, Blanche Demagall, the Ziemba Family, Bob and Janice Veronesi, Phil Leahey and Jen Dustin, Ted Jayko, Flossy Hollerich, Dave and Georgine Long, David and Doris Alibozek, Murray Hochman and Lois Jensen, Ann Barrett, Patricia Clark, Peter and Tjasa Sprague, Wanda Kieltyka, the Haddala family, Jean Paul Blachere, Harold and Doris Shaw, Mary Beth and James Ivanowicz, Sean Stanton, Roberto Lawrence, Johnathan Wilk, Will Garrison and The Trustees of Reservations; Gary LeVeille and James Parish of the Great Barrington Historical Society; Nancy Hahn, Sue Young, and the Bushnell Sage Library.

There are many other people whose barns I photographed from a distance and whom I did not have the opportunity to meet. I want to thank them for understanding that it is not always possible to make those connections when undertaking a project of this scale. I would also like to express my appreciation and gratitude to anyone who has embraced the cause of saving and preserving barns wherever they are located, and I hope that as many barns as possible will still be standing in 100 years time to provide a proud and rich visual testimony to the early history of our country for generations to come.

Once again I thank my wife, Sarah, for all of the good she brings to my life; my friends, family and extended family for their unwavering encouragement and support; and, well, to Marquis, my cat and silent partner, the one and only Quartermaster of the Poop Deck, whose companionship I have enjoyed for all but one-and-a-half of the last eighteen years—I love you man. Here's to another successfully completed project—now it's on to the next one.

Left:
The substantial barn complex at Lila's Sheep, also known as Stone Hedge Farm on Seekonk Cross Road in Great Barrington, is spread across a large portion of the hillside. There are five barns, with the oldest, shown here, dating to 1790, and several sheds. Owners Peter and Lila Berle invested in extensive maintenance, repair and restoration work on the farm's building for well over 20 years. The magnificent gable roofed red barn, built in 1860, that stands above all of the others, rests on this enormous stone foundation. Poor restoration work done several years ago caused the wall to begin to collapse so the whole wall, almost five feet thick at its base, was rebuilt in 2007 ensuring that it will remain standing for at least another 150 years.

Preface

Preparing and researching this book has opened up a whole new world of exploration for me. At its core this project has involved a study in architectural development and styles, which in turn brought to me a deeper understanding of the demographic, sociological, and agricultural history of the Berkshires, and much of the northeast United States for that matter. A more profound knowledge of the countless whys, whens, hows, and wherefores of barns has made me appreciate and treasure them far more that I ever did.

I have also come to appreciate how daunting a task it can be to try to piece together history for which there is very little written record. Barns are buildings of function and practicality. For 600-700 years they were simply a necessity without aesthetic implications. Unlike the pyramids, the Acropolis, the Colosseum, or Notre Dame Cathedral, they have not been revered and studied, either individually or collectively, by historians, archeologists, and sociologists for centuries. But their impact on human life on earth has been far more essential and profound than that of any of these structures. For the first 150 years of development in the American northeast and New England there is scant documentation of barns—no architectural plans or drawings, no novels or classic, multi-million dollar landscape paintings to go by.

Much of what began to emerge in the study of New England barns was theory based on logic, practicality, and the knowledge of barn building techniques acquired on the job. All of the sources I have drawn upon derive from earnest, intelligent people who worked hard and cared deeply about barns. Over the years further investigation has yielded new information, and altered theories and accepted truths, but it has not diminished the profound value of the work and the knowledge that these people contributed to the evolution of our understanding of barns. The work they performed presented many pitfalls. Barns were never meant to be permanent monuments to the prowess of the human race. They had an inherently tenuous life; they were ultimately bound for obsolescence and they were always in danger of being destroyed by fire. Many barns were moved one or several times, some had to be rebuilt because of rot or poor original construction, some were enlarged over time. After a fire a farmer would often recover whatever wood he could from the barn he had lost and incorporate these timbers in a new building. It was very easy for casual or untrained observers to believe that the newer barn, a hybrid of old and new, was the original barn. For researchers this often meant carefully differentiating between an original and a next-generation barn. Many owners of barns have falsely believed that they possess barns that are much older than their essential parts and construct might suggest.

I am neither craftsman, carpenter, historian, or sociologist—just a photographer—and so I am humbled by, and deeply respectful of, all those whose dedicated work in these fields has made it possible for me to produce this book. My goal was to combine and condense what I learned from their work to produce a book which is informative, factual, and accurate while also being interesting and engaging for all audiences.

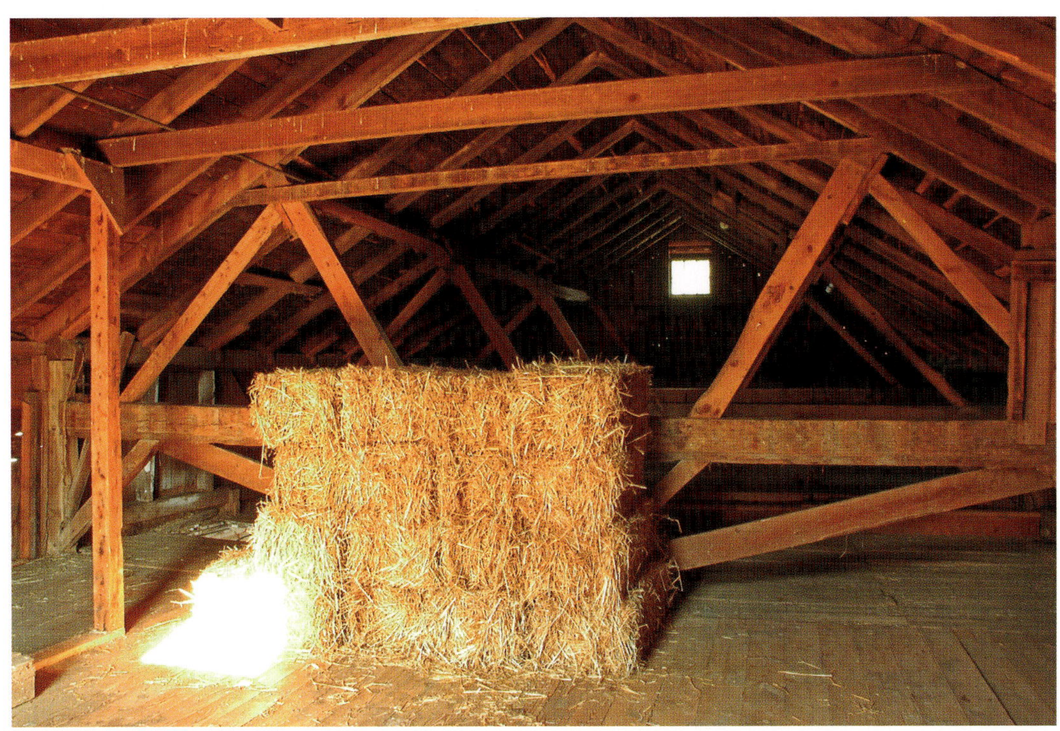

Roof structure of Ron Majdalany's English gable-entry barn, c. 1807, on Alford Road in Great Barrington. The barn has a non-operating milking area with concrete floors in its southern half; the northern half has been converted to house a veterinary hospital.

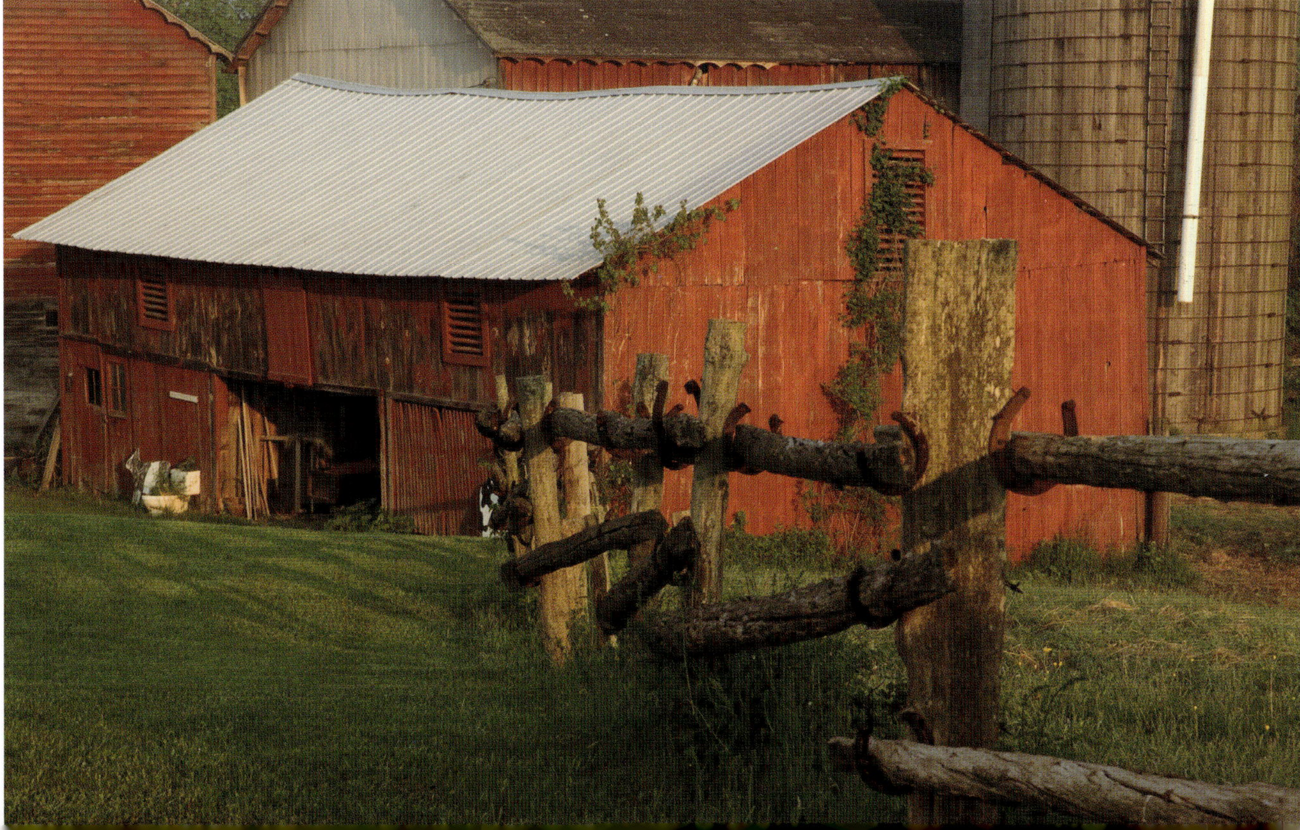

Side-entry pole barn at Delmolino's Farm, Sheffield, c. 1900. This barn, built around the same time as the main barn, is part of a complex of six buildings clustered around the original farmhouse that once served as a stagecoach stop. The unique horseshoe fence was designed and built by Deborah Storey with horseshoes found on the property which the Delmolino family bought, fully stocked, in 1924. There were two horses on the farm at that time. When they came up a little short, family friend Noel Alsop donated a few horseshoes from his stables.

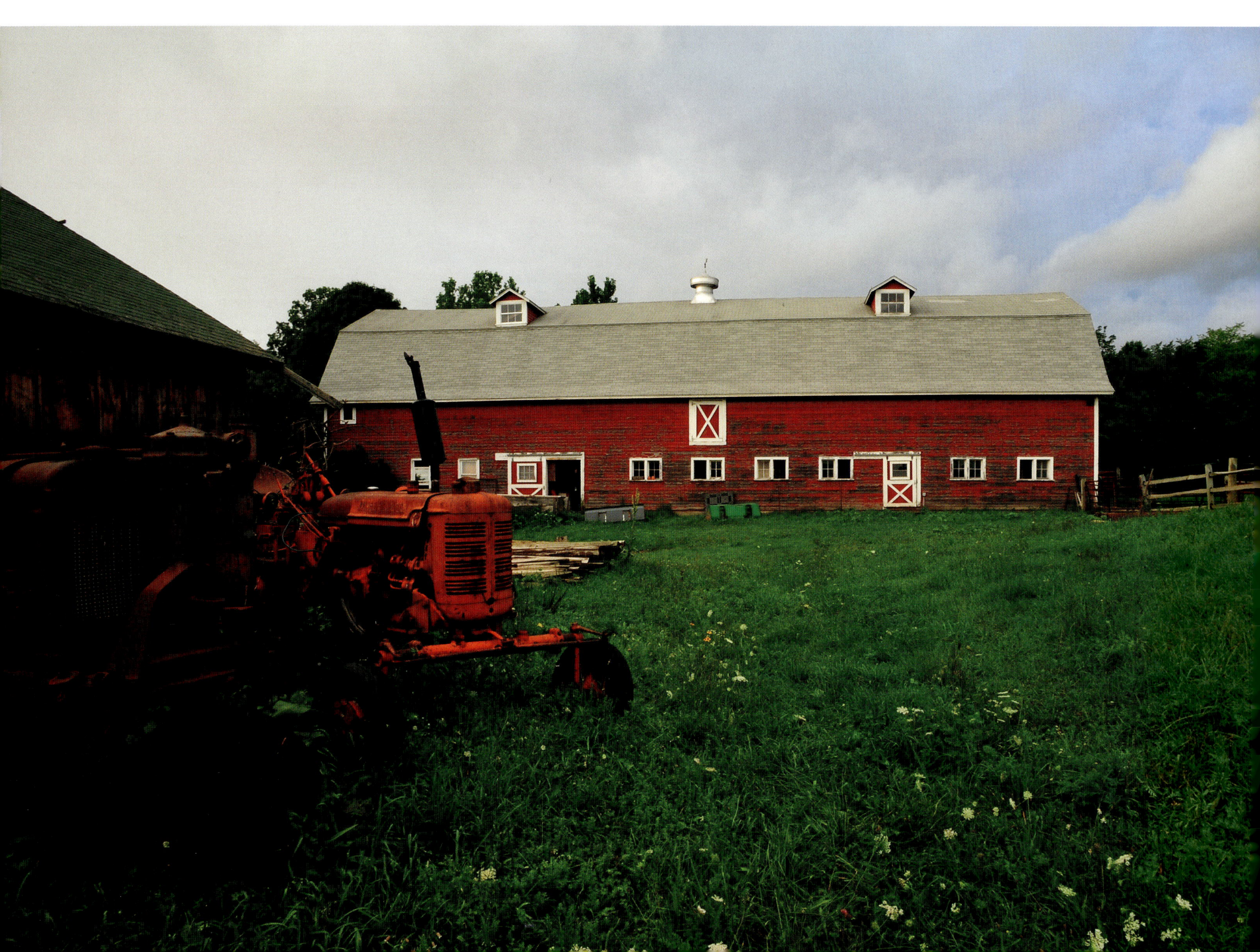

Introduction

Like many of the readers of this book, I suppose, I have had a quiet fascination for barns for as long as I can remember. It wasn't like a young boy's absolute love affair with trucks, or bulldozers, or trains, I just always liked and appreciated barns whenever I saw them, and enjoyed exploring them if I had the chance.

Part of this, I am sure, derives from the fact that I started out my life as a suburban kid surrounded by well-tended, orderly residential streets not far from what was at the time, one of America's great cities. Every summer of my first seven years my family would pack up and head north out of Detroit, into the lower regions of Michigan's thumb, across the Blue Water Bridge, and into the southwestern appendage of Ontario that provides Canada's coastlines with Lake Erie and Lake Huron, en route to my grandfather's cottage in Grand Bend, Canada. Once beyond the fringes of Detroit, which, at the time, were demarcated by the convergence of Gratiot, Little Mack, and Thirteen Mile Roads, we were in farm country.

From the backward-facing rumble seat of a town-and-country station wagon my older brother and I would alternately irritate each other, wave at other drivers, and gaze out at the surrounding corn fields. There were barns galore, and while we didn't make any kind of fuss over them, my mind was busily making note of them. Since I knew absolutely nothing about barns, I didn't differentiate between the various styles, shapes and sizes for the most part, but I've always had the impression that the farms on the Canadian side were larger operations with bigger, more impressive barns.

My awareness of barns became much more acute when, in 1969, my family moved to England. For the next seven years we lived in a "green belt" between Coventry and Kenilworth surrounded by quintessential rolling English countryside. Within a three-to-four mile radius we could access, among other things, the Royal Agricultural Center in Stoneleigh, the mystical ruins of one of England's great medieval castles, and the spot where Lady Godiva bared it all on horseback nearly one thousand years ago.

One of my first best friends was Nicky Gee, the gentle son of a gentleman farmer. His father owned a dairy farm on a hillside above the sprawling grounds of Kenilworth Castle. There were several barn buildings on their property, but I remember most clearly one cavernous, musty old hay barn where we would play. It is the first time in my life that I can remember experiencing the thrill of jumping twenty feet from a beam into hay and building forts in the dark upper reaches of a loft. I only knew Nicky for a few years before he decided to leave the school that we attended, but I have cherished those memories of his family's farm for the last forty years.

Left:
The main barn at Leahey Farm in Lee burned down in 1934 and was quickly replaced with this beautiful gambrel roof barn. The replacement is a 100 ft. by 30 ft. Starline barn. It has stalls for four draught horses at the western entry, a drive-through loading bay, and two roughly 60 ft. rows of stalls in the rear eastern end that accommodate up to 34 head of cattle. The exterior of the barn is finished with tongue-and-groove novelty siding that had become popular in the 20th century.

In that seven year period my family traveled a great deal throughout the United Kingdom so it would have been difficult not to have developed an awareness of the ubiquitous, centuries-old symbols of agrarian life that filled in the landscape between ancient towns and villages. We visited countless storied estates and occasionally took in one of the vast tithe barns, built in the shape, form, and scale of a great cathedral. They were, for hundreds of years, the storage sites for the tithes, or taxes, of Britain's feudal serfs. And, again, while none of these experiences led me to any sort of life-changing epiphanies, they did form an indelible image of these great structures in my mind.

When we returned to Detroit in 1975, my life resumed its urban-centric character and would stay that way through stints in Detroit, New York, and Los Angeles for the ensuing twenty years. The more I lived in cities, however, the more I enjoyed my time away from them. I always cherished my trips back to Canada each summer, and whenever I had the opportunity to escape, for any length of time, from the chaos and congestion of either New York or Los Angeles it was most often to open country. During the eight years that I lived in New York I took many trips north to Vermont, and from Los Angeles I was constantly drawn to the magnificently diverse National Parks that dominated the city's hinterlands.

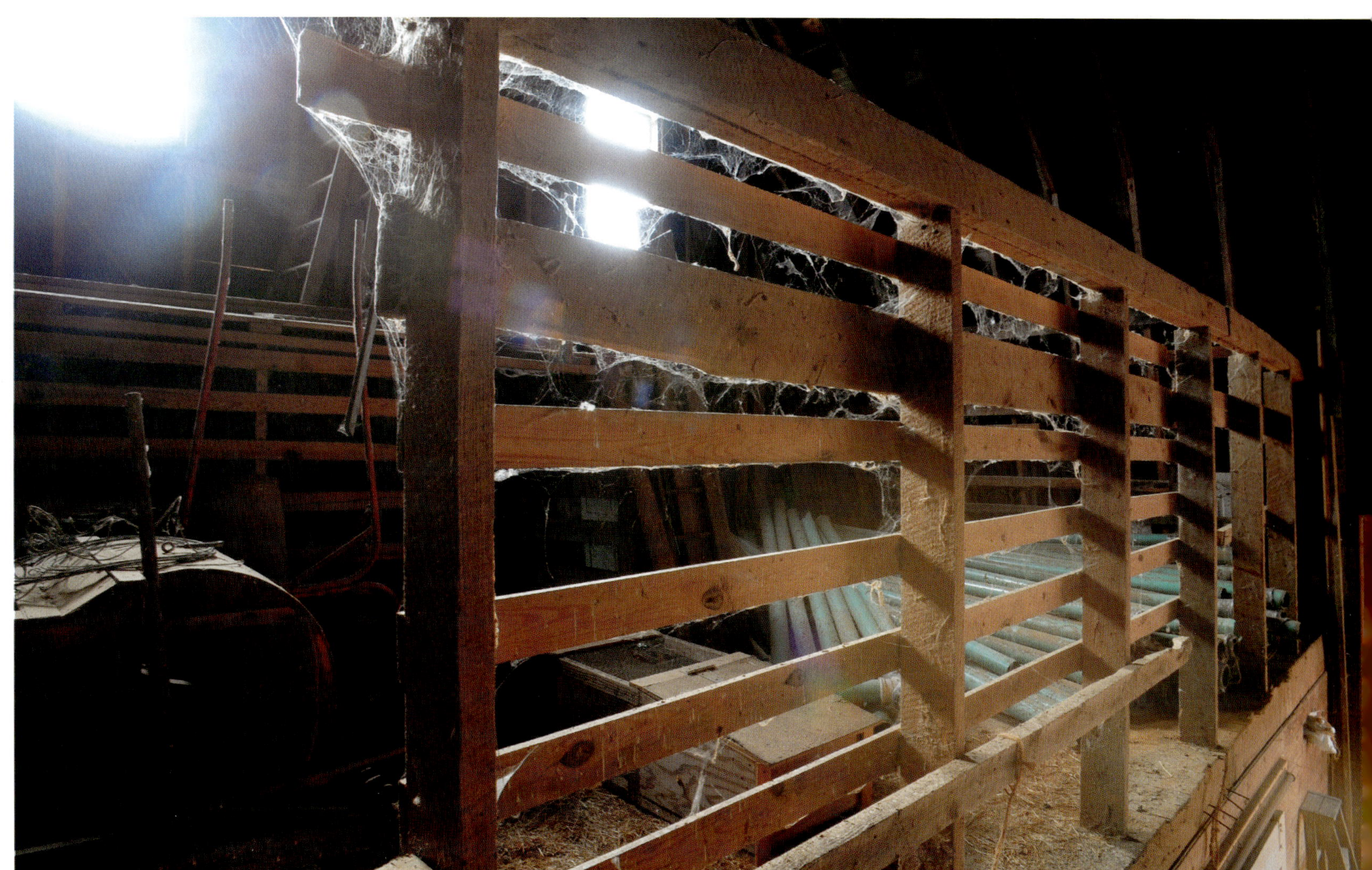

Late-day light streams into the loft area of the Leahey barn.

It seems fitting, then, that the place where I would eventually settle and where I would become a home-owner for the first time in my life, would be a pastoral, bucolic corner of New England where I have discovered so much that is reminiscent of my childhood in Coventry. I came here after having left a series of corporate jobs to pursue a career in photography. Almost from the beginning my interest in barns was rekindled as I explored the back roads of Berkshire County with a camera or two at my side. Yet it wasn't until I undertook this project that I discovered how significantly related the barns that I have photographed for the last ten years are to the place where I grew up. It was always easy to recognize the physical similarities of the landscape, but it was only through researching the lineage of Berkshire County's barns that I learned of their uniquely English heritage.

And so, as I have dug deeper into the subject, my fascination with barns has morphed into a love affair. I have become attached to many of them and have mourned the loss of more than I wish to account for. There are a number of barns in this book that I have photographed on multiple occasions and in all of the seasons. I have gotten up at ridiculous times in the early morning hours and driven as far as 50 miles away in the pitch black to photograph them when I thought the sunrise light would be just right, or waited until the last minutes of daylight to try and capture a new facet of one in the rich, red glow of sunset. Sometimes I succeeded; often I failed; but the buildings always seemed willing to oblige if it were in their hands to create the effect I was looking for.

Frequently, as I suggested above, I have been made painfully aware of the fact that our barns are an endangered species, albeit an inanimate one. They are dying off at an accelerating pace, and not just in the Berkshires. All over the country people are hearing more and more about the disappearance of barns. In 2008 the Vermont Division of Historical Preservation announced an effort to catalog all of the state's barns in a statewide, volunteer-driven barn census. There have been some published estimates that as many as 1,000 barns disappear from the Vermont landscape every year, and the statistics are similar in Massachusetts and the other New England states. Most of these, of course, are the older historic barns that provide the scenic character that is so sought-after here.

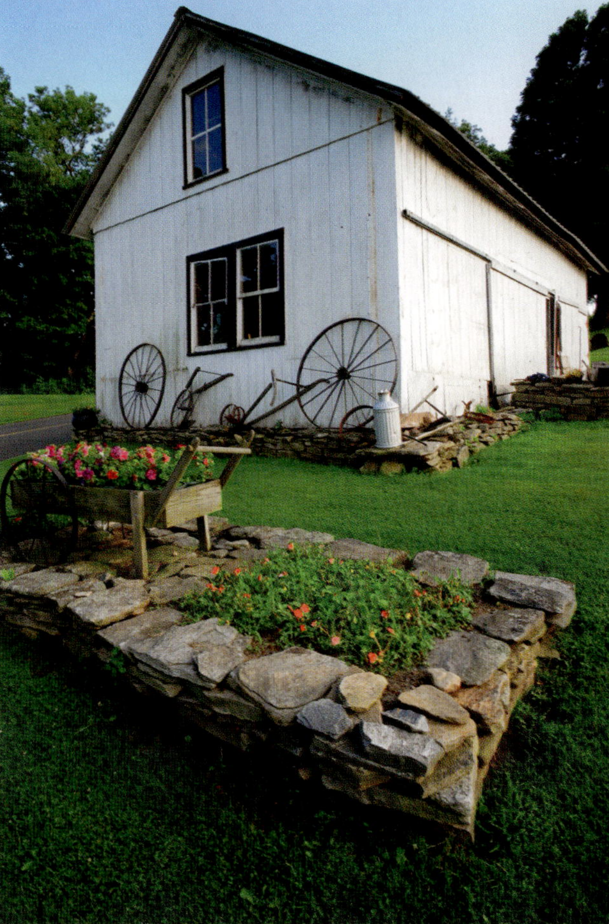

A Shaker barn, c. 1850, was relocated from the community at Fernside down to the village of Tyringham and now serves as a garage.

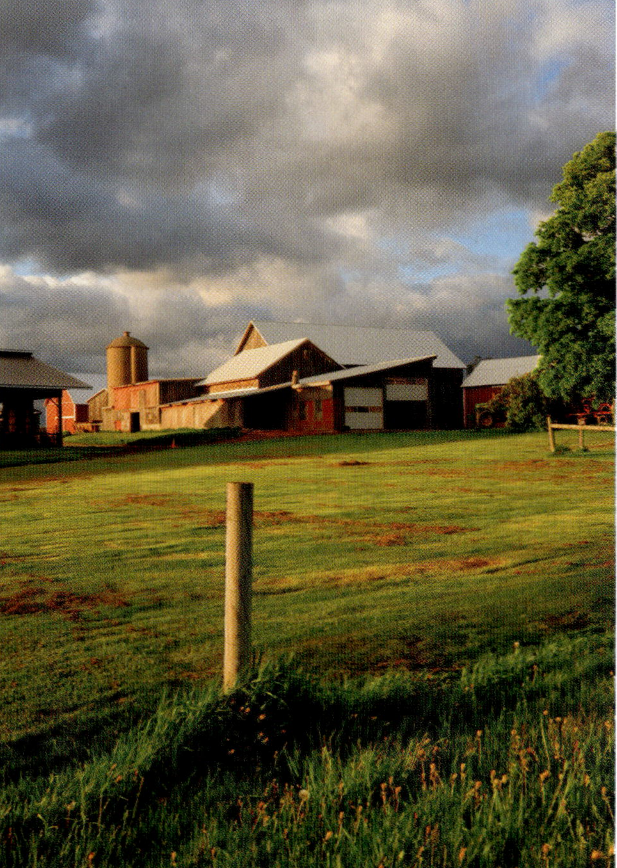

Sunset at Proctor's Farm on Baldwin Hill in North Egremont.

Since taking on this project I have also discovered how much of a body of work exists about barns. It seems that awareness about the looming fate of historic barns began to build about 50 years ago. Much of the work done on the topic did not achieve a high profile in the public's eye and was limited to local Historical Societies and to the efforts of barn builders like Richard Babcock who had a great reverence for the structures and the craftsmanship that went into their construction.

Babcock had been taught how to hew timber and frame barn structures by his grandfather. He learned the nuances of mortise and tenon joinery and was one of the early pioneers of deconstructing and relocating historic barns in the eastern United States. He chose to erect his barns with the tools and technology used by farmers and builders two hundred years ago—derricks, gin poles, pulley systems, and pike poles. He became devoted to his dream of preserving a collection of prototypical New England barns in a museum on his property in Hancock, Massachusetts, until a devastating fire extinguished any hope of achieving it.

In the 1960s social historian and illustrator Eric Sloane authored numerous books on barns and barn building, including *An Age of Barns*; in 1984 Thomas Hubka published his frequently-quoted book *Big House, Little House, Back House, Barn*; Thomas Durant Visser released *Field Guide to New England Barns and Farm Buildings* in 1997; also in 1997, Randy Leffingwell produced a beautifully illustrated book titled *The American Barn*; the Trustees of Reservations published the concise *Barns in the Highland Communities* which covered Hampshire, Franklin and Berkshire Counties in Western Massachusetts; and Ron Majdalany, a veterinarian in Great Barrington, even self-published the 34-page *The History of Seekonk Farm* in 2002 chronicling the life of his barns since their 1806 origins. These are just some of the dozens of books that can be found on the subject of barns. The amount of material published about barns seems to have accelerated in the last 25 years, and so, like so many other things, people didn't really begin to appreciate them until there was a collective realization that they were losing them—until they became an endangered species.

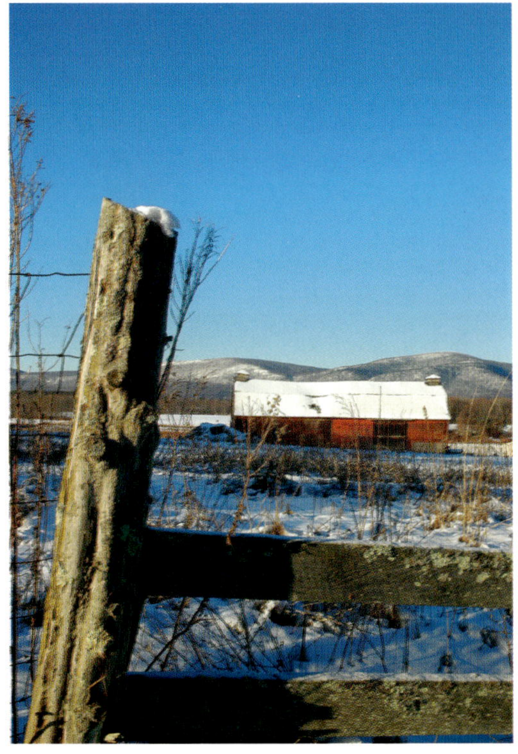

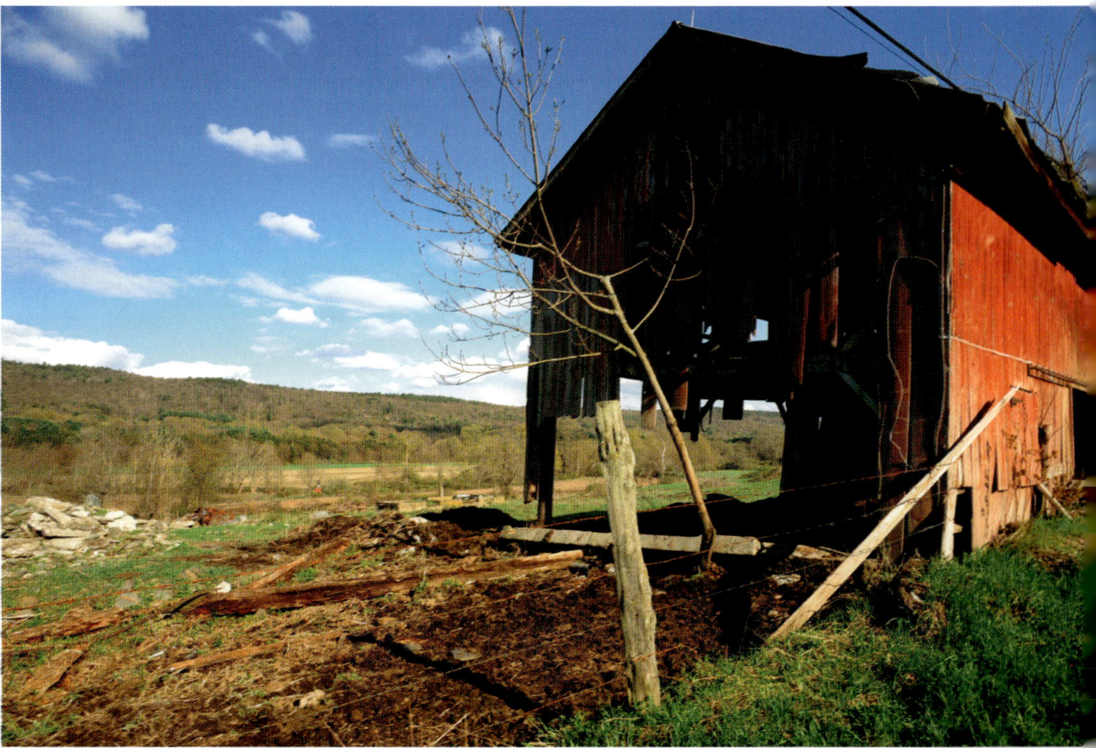

The huge barn at Boardman's Farm, seen here from Hewins Street with Mount Everett State Reservation in the background, labors through another hard New England winter. It is one of the endangered species. The roof of the barn has begun to collapse and it has little use beyond equipment storage. According to Boardman, maintenance and repair costs for a structure its size are prohibitive, and he has no choice but to let it suffer a noble and gradual death.

To date, however, most of the works published about barns have not been as regionally specific as what I set out to accomplish. In fact, aside from the booklet produced by The Trustees of Reservations, there is rarely a reference to any barn in Berkshire County in the books with broad geographic coverage. For this book I wanted to show what intrinsically valuable assets the barns of Berkshire County are, and how essential they are to the visual fabric of the region that makes it such an appealing place for both visitors and those seeking an alternate lifestyle to the charge-ahead pace of our cities. For me the barns are symbols of integrity, wholesomeness, and a down-to-earth ethic that was an essential building block for this country.

Barns are also the great symbolic icons of past and present rural life in America. They have a permanent and compelling place in the hearts of all who have traveled the North American countryside and observed them. Yet, despite this growing awareness I have outlined above, there seems to be very little that has been accomplished to preserve barns in their native landscape. Of course, although they may, in one sense, be public treasures, they are not public property and the cruel financial realities with which most farmers must contend make barn preservation a low priority. The utility of most pre-1850 barns has completely diminished because they cannot accommodate the overall scale of modern farm operations or the size of the equipment. In many cases a farmer's only option has been to sell older barn structures directly to private buyers or to spec-builders who dismantle and relocate them to build custom post-and-beam private residences and other structures. Wealthy individuals from places like California, Idaho, Montana, Boston, New York and the Hamptons have ponied up as much as $500,000, and more, to relocate New England barns for their private use as homes, studios, and even swimming pool pavilions.[1] And while there is some consolation in the fact that the barn does not simply rot and disintegrate, the fact is that it will no longer be seen in its original context, if it is ever seen by the general public again at all.

Left:
This old side-entry English barn on Gary Shepard's dairy farm on West Alford Road is one of the doomed barns that will probably have disappeared in the next few years. Beyond repair, and with probably little salvage value, it stands over Alford Valley like a hulking shipwreck.

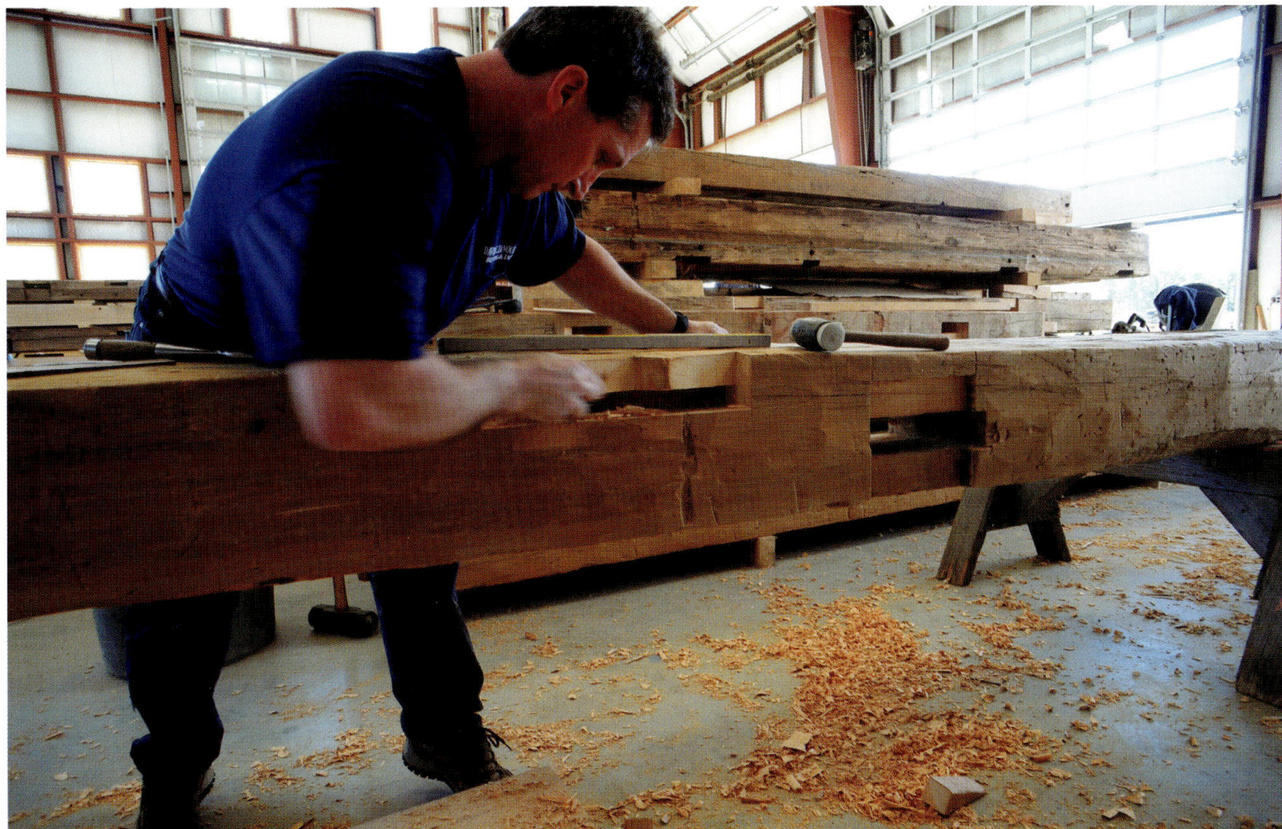

One of the master carpenters at David E. Lanoue Inc. chisels a mortise into an old hewn timber of a swing beam barn being relocated from Ontario, Canada to the Berkshires. Old barns flow in both directions in the restoration trade, but the Berkshires have had more buildings being brought into the area since 2000.

Occasionally, as the *Christian Science Monitor* reported in February, 2003, whole communities have responded to specific cases of treasured barns that were targeted for relocation. In that issue, Abraham McLaughlin reported that the residents of Penacook, New Hampshire, lobbied the city council to invoke eminent domain to prevent the relocation of the "Rolfe Barn" (c.1880) which had been purchased by a local restoration specialist who had then contracted to reconstitute it for a buyer from a western state.[2] The story gained national exposure and in the end the town prevailed. The barn now serves as a centerpiece for an exhibit and meeting complex under the stewardship of the Penacook Historical Society. Similarly, but with much less fanfare, the Great Barrington Historical Society, here in the Berkshires, successfully rescued the Truman Wheeler Farm, comprised of six historic barns and a farmhouse dating to 1766, from potential commercial development.

The essential irony of all of this, for which Berkshire County is a prime example, is that ongoing and increasing development of rural areas is driven, in large part, by an urge to flee urban settings and lifestyles for a more subdued environment filled with rolling hills and cow pastures and barns—I suppose you could call it the Green Acres Syndrome. It certainly isn't a new phenomenon, but its appeal has grown at an exponential pace in the last 10-15 years. But, the more people there are that seek this ideal, the more the land is consumed and permanently altered so that the romantic landscape that drew people here in the first place gradually fades from sight. With it go the barns and the farms.

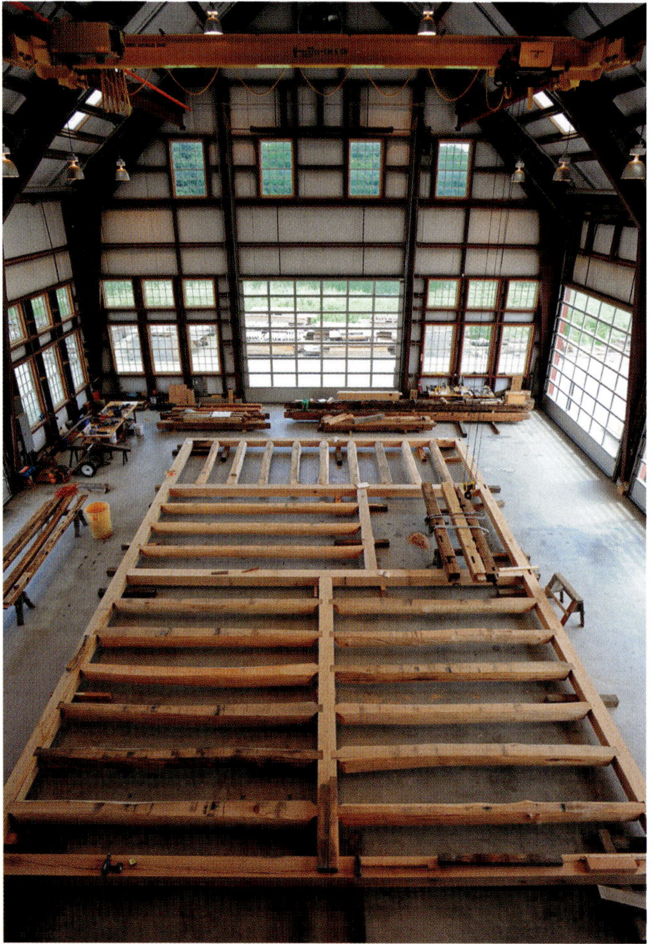

The framing for the floor of an English three-bay barn dating to 1828 occupies almost the entire work area of Lanoue's state-of-the-art facility on Van Deusenville Road in Great Barrington. Eventually the barn will be completely erected inside the facility to mark all joints and check for fit before it is dismantled and taken to its new site.

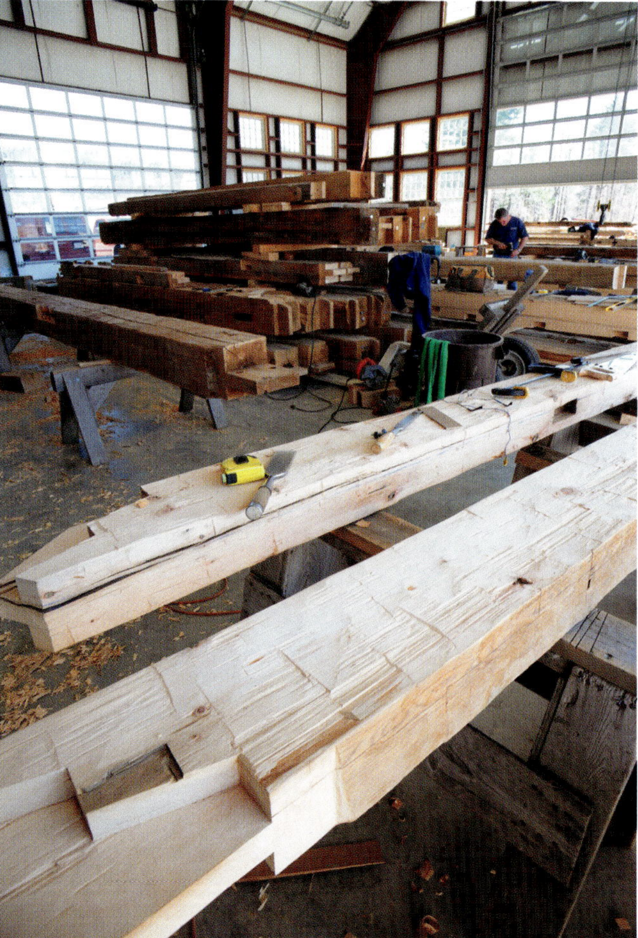

Timbers being prepared for assembly.

My hope, in doing the work I have done for this book, and in other associated published works, is to bring awareness to the notion that we need to muster support from outside the farming community to assist farmers and other land owners in the preservation of these great buildings. This book contains over one hundred images of more than sixty individual barn structures. In every instance my motivation for photographing a barn was to capture the way in which the barn either enhanced or defined the landscape in a beautifully nostalgic way. At last count two of the barns that appear in this book have completely vanished from those landscapes, and another six are in such disrepair that their demise is imminent. In every case, the landscape has become significantly less attractive and appealing. Without some sort of concerted and concentrated effort this will become the overwhelming and irreversible reality for the vast majority of the barns of Berkshire County that you will see in this book.

Finally, while it may be obvious, this is not a book about every barn in Berkshire County, and there is no possible way that I could have included every barn in it. There will, however, be some people out there who feel that the book is an incomplete record because their barn, or one that they know and admire, does not appear in it. My only response to this is that I have, to the best of my ability, scoured Berkshire County in an attempt to compile a comprehensive record of as many barns as I could find that possessed aesthetic visual qualities *and* that I was able to photograph at the right time, and in the right conditions, so that these visual qualities would be revealed to the reader of this book. Some people specifically requested that their barns not be photographed, others were simply not home when I happened by and knocked on the door to ask permission to take a picture or two.

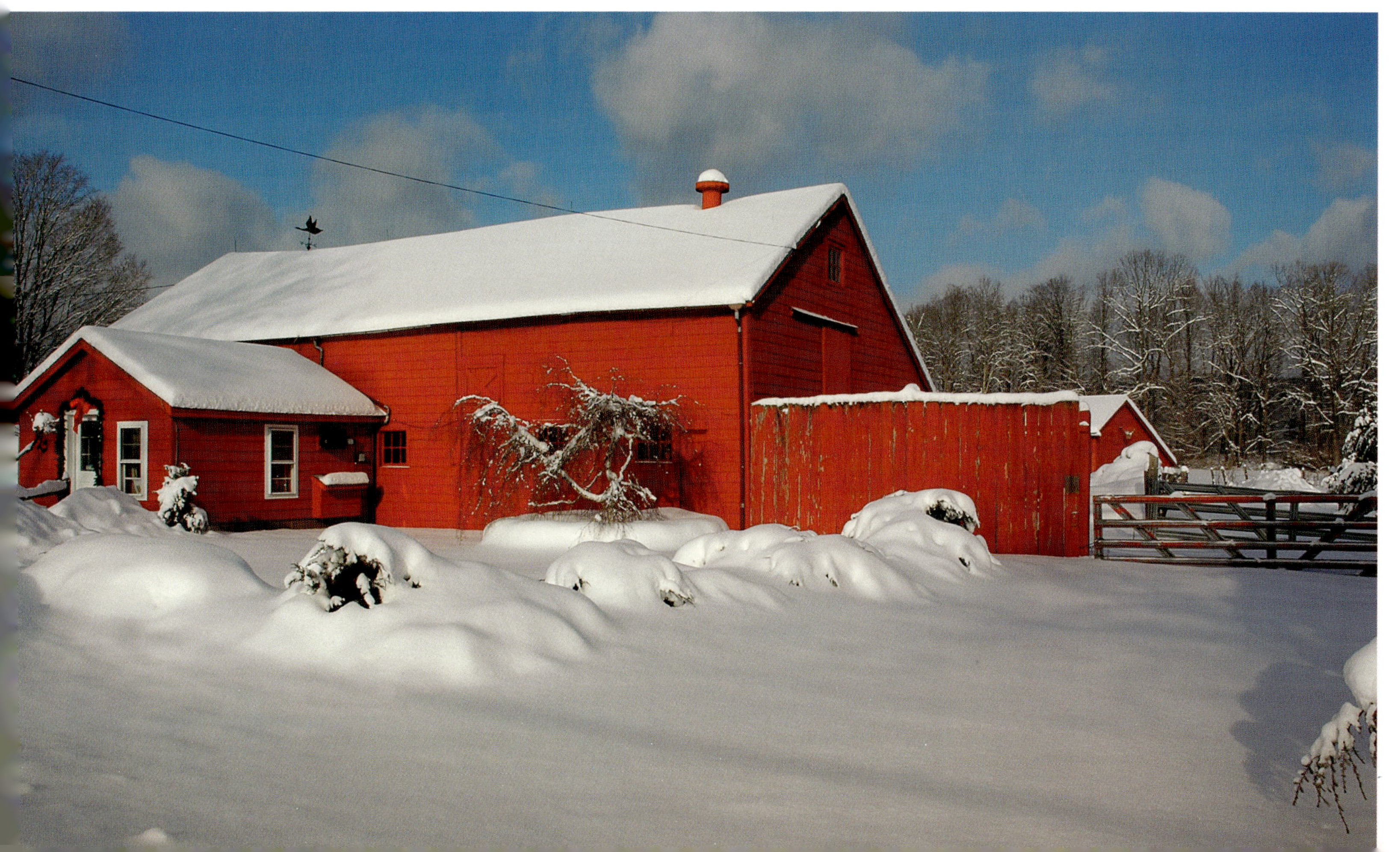

In 1983 Ron Majdalany, fresh out of veterinary school, had a golden opportunity to realize several of his goals at one time. He wanted to open a veterinary practice in Great Barrington fashioned after one he had recently become acquainted with in Sharon, Connecticut, that had been set up in a former dairy barn. He knew that a farm where he had often worked in his youth, and for which he had a strong affinity, had been put up for sale, due to a series of unfortunate circumstances. Despite some initial setbacks, Majdalany was able to purchase the farm along with this 1807 gable-entry barn and his practice has operated there ever since.

Wherever possible I have provided dates for the approximate original construction of a barn. In many cases, however, I simply photographed a beautiful barn that I happened across at the right time and I have no other connection with it, or information about it. I call these my drive-by shootings, and they are intended to be completely harmless. This is, above all, a photography book, and so, while it does not stand as a reference book of anything near encyclopedic scope or quality in the text provided, I sincerely hope that every reader will find richness and beauty in the pictures and be moved in some way by the final product that is laid out in these pages.

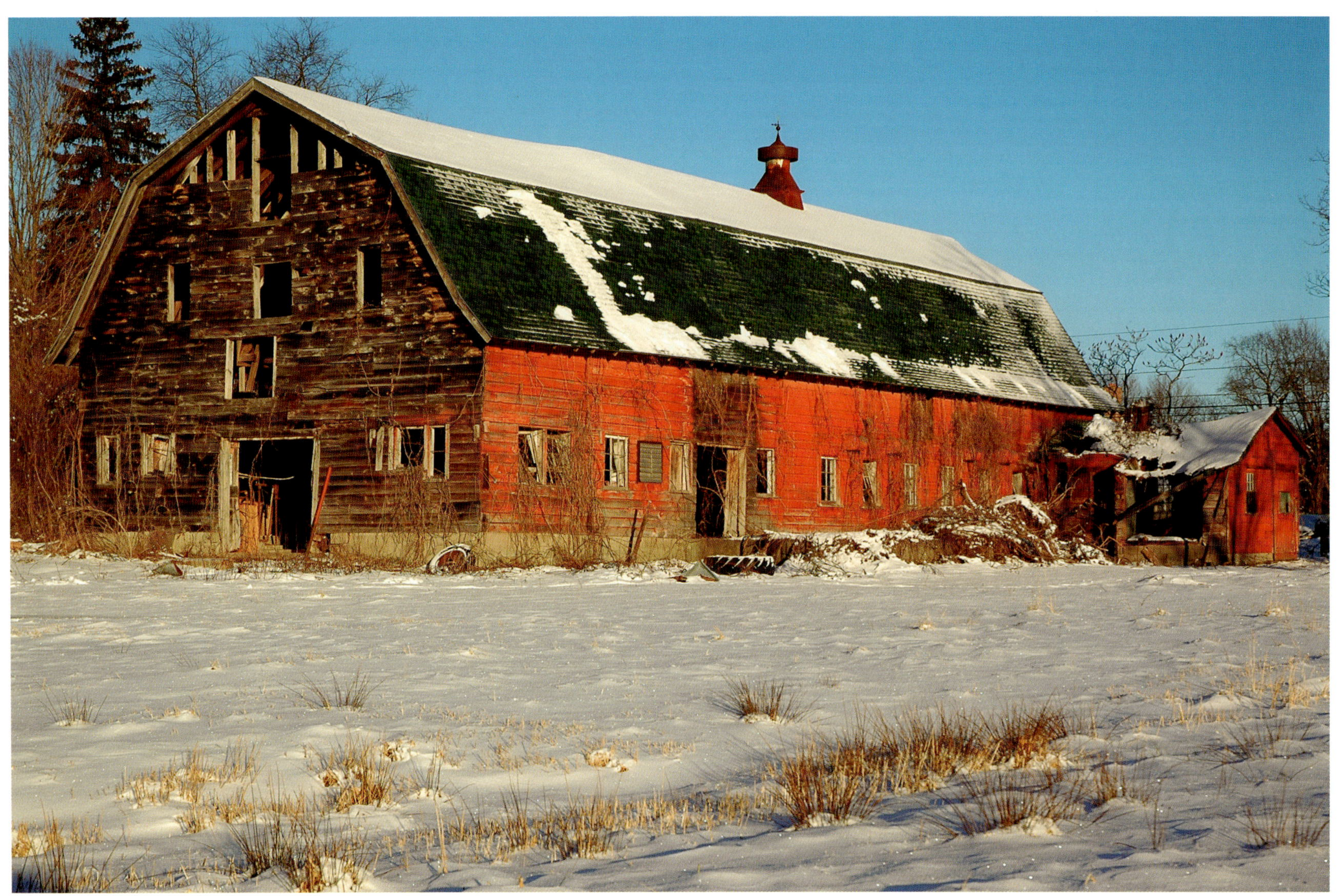

This large gambrel-roofed dairy barn in Sheffield is on land that is for sale and it is probably destined for the wrecking ball. Owners and date of construction unknown.

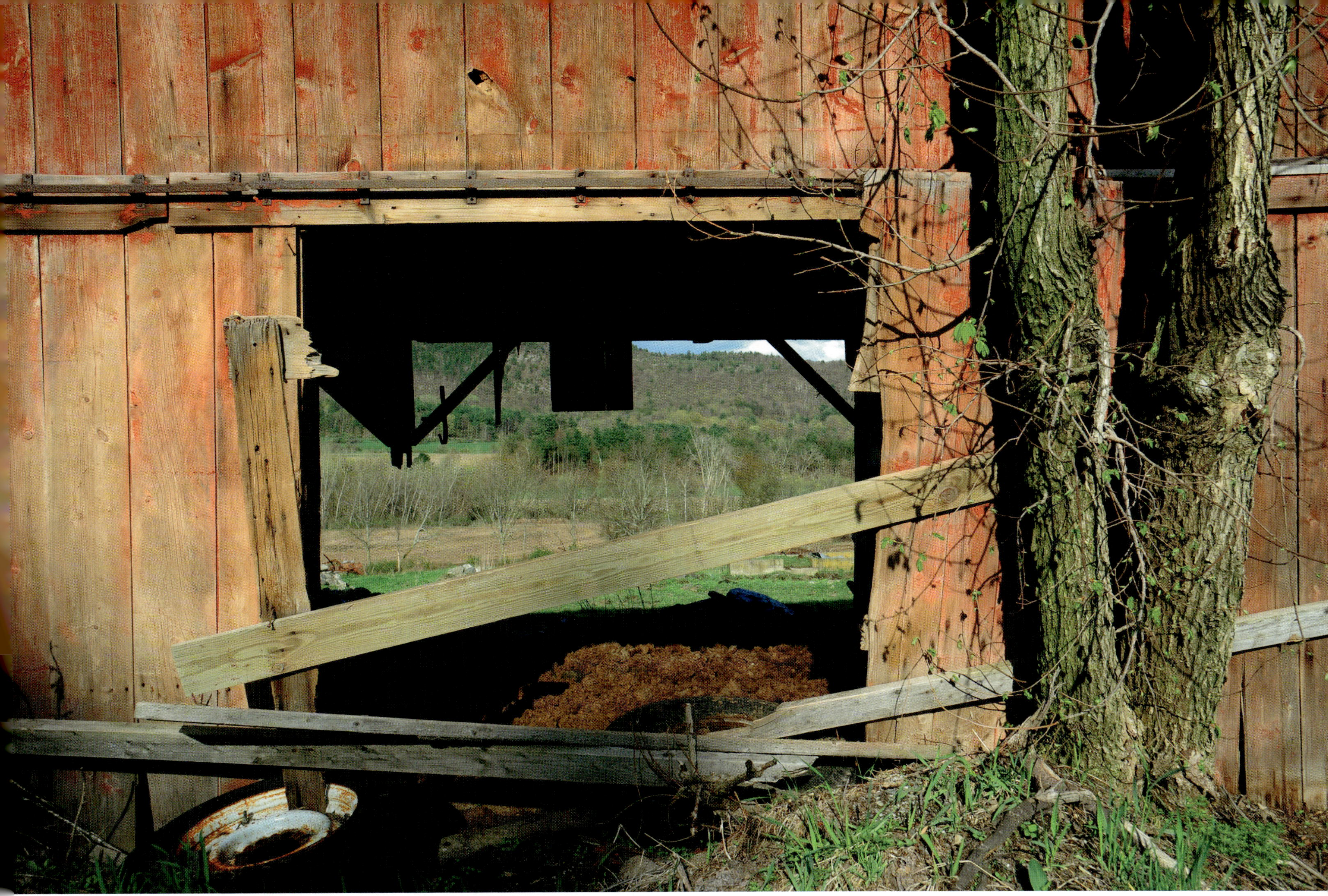

The view through the dilapidated walls of Gary Shepard's barn in Alford reveals a commanding view of Alford valley that it has claimed for at least 100 years.

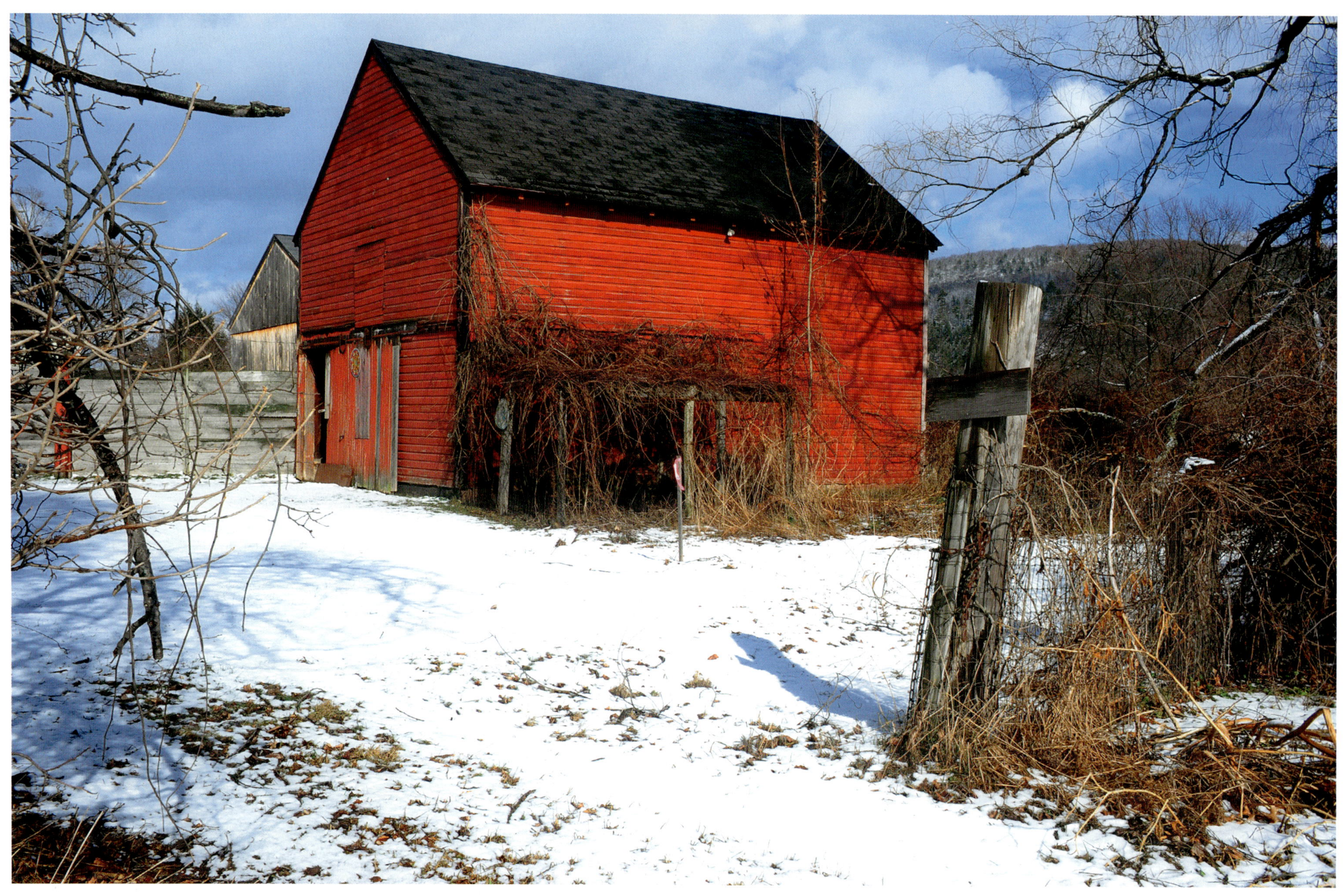

Part of the Truman Wheeler homestead and farm that was purchased by the Great Barrington Historical Society in 2007 as its future home, this c. 1825 gable entry barn is still in remarkably good condition and is the most noticeable feature of the property to passersby.

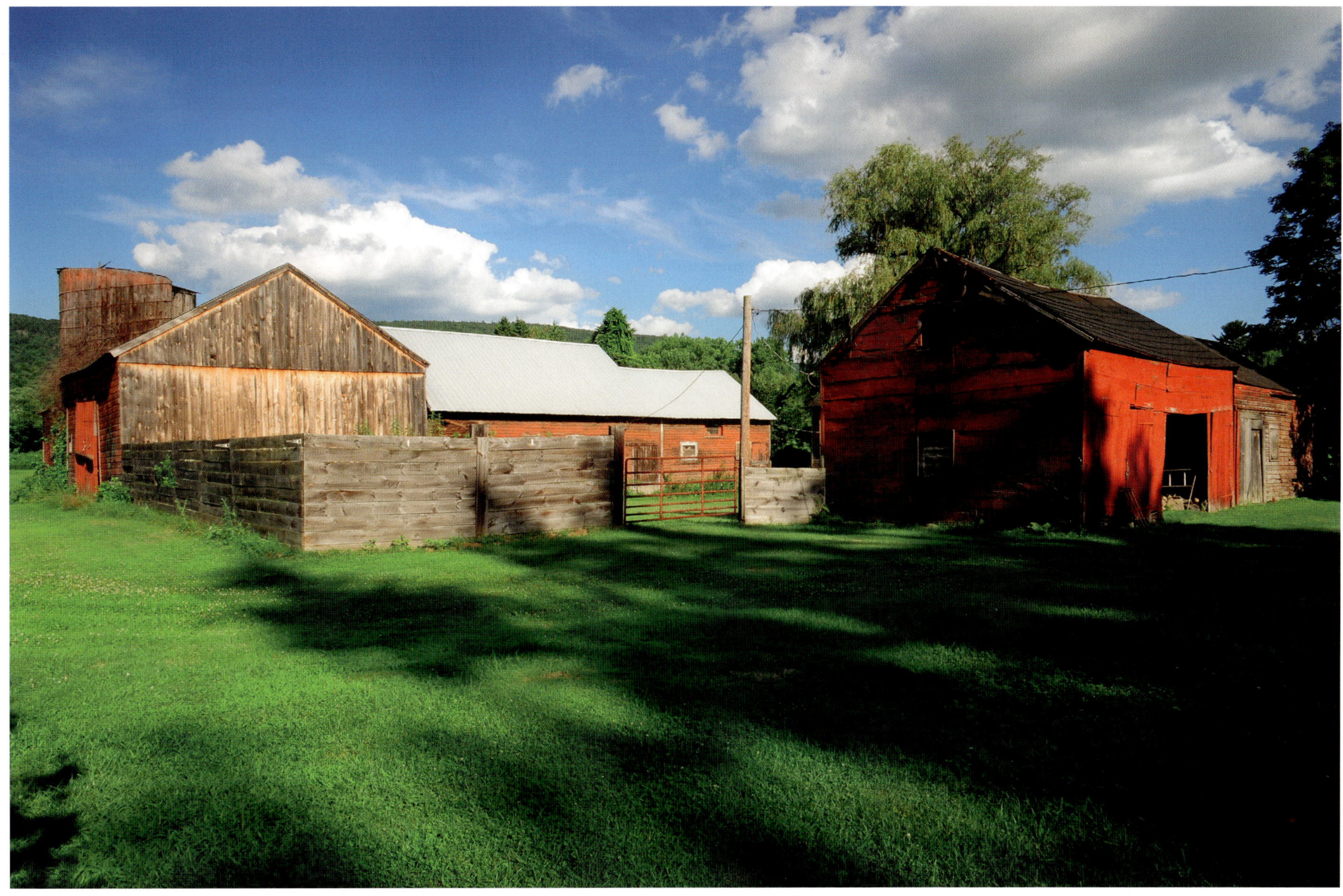

The barn complex of the Captain Truman Wheeler House viewed from the northwest side. This 1.5-acre property is now just a fraction in size of what it used to be, but the essential buildings—six barns (including one of the earliest surviving barns in Berkshire County, c.1771) and the main farmhouse dating to 1766—are destined to remain intact as a museum. The property is truly a community treasure; it was the home of one of Great Barrington's pioneering settlers and has links to the Revolutionary War and the early economic foundations of the town. Descendents of Wheeler were involved in the sale of the property and passed up offers of significantly more money from commercial developers because they could not stand the thought of losing it forever. As of late 2008 the Historical Society was managing to keep the whole operation moving forward, but was seriously in need of more broad community financial support.

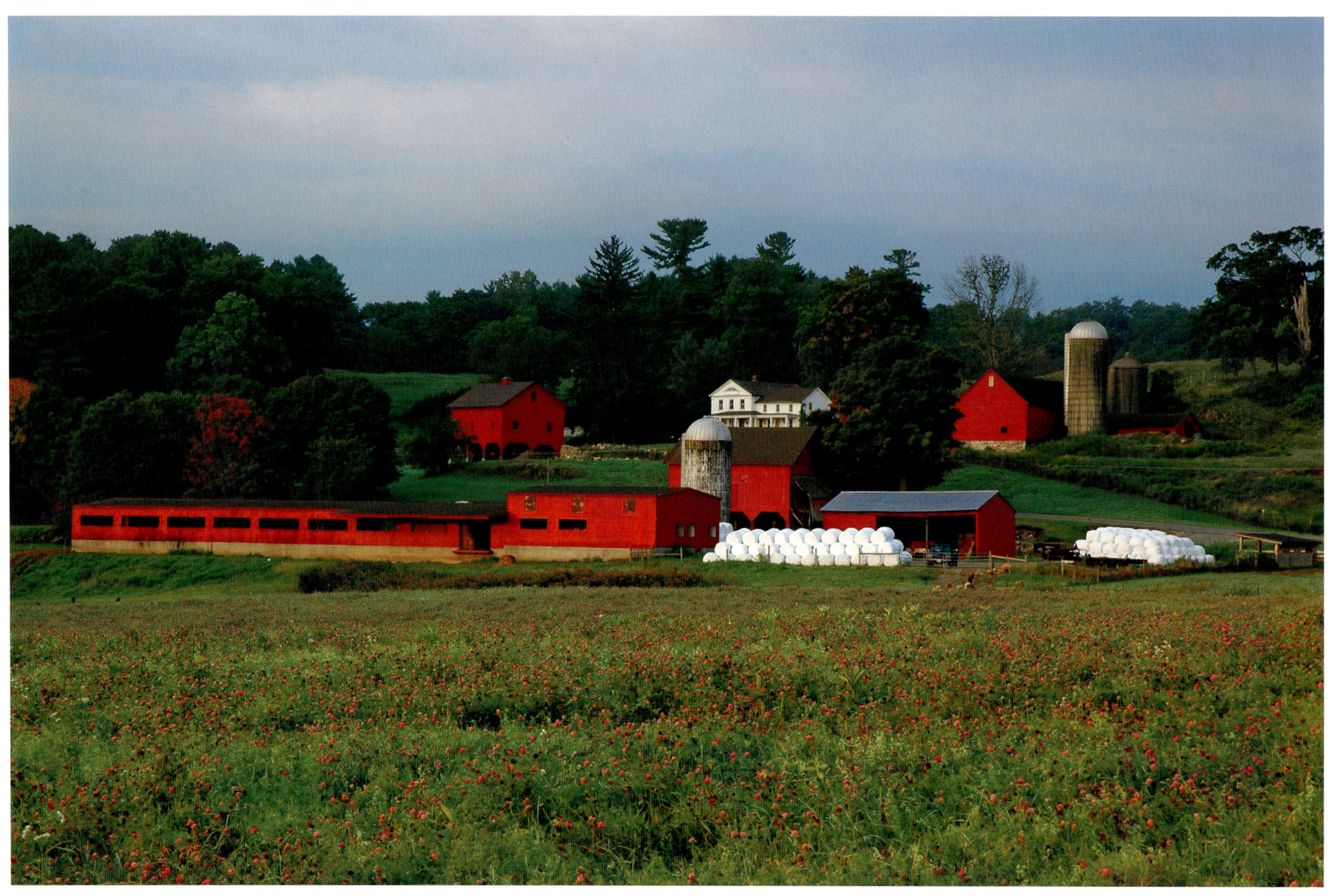

The sheep farm on Seekonk Cross Road is truly one of the most picturesque spots in all of the Berkshires. There are many approaches by road, each offering different, but equally compelling, views of the farm and the surrounding mountains and countryside.

The Pre-History of Berkshire Barns

The word barn seems to have surfaced sometime in the late 1500s or early 1600s. It originally meant "a place for barley" and was derived from two Old English words—bere, meaning barley, and ærn, or ern, meaning place or house. A variety of early spellings have been found that include beren, bern, and barne.

The earliest form of Old World barn, from over 1000 years ago (and probably derived from Greek and Roman practices), was little more than a pit for crop storage dug into the ground and covered with some rudimentary form of hatchway or door. Other variations would have been simple ground-level, A-frame, tent-like structures of straight poles covered with a layer of hay. Recent discoveries have shown that framed barns were beginning to appear in England in the 12th century. Most examples of barns dating to this period employed two crucks, or curved beams, in each section of framing which created an arched appearance inside the building. There are a number of theories for why the carpenters of the time chose to employ this framing style. Some suggest that it was because of a dearth of tall, straight lumber. Others believe that builders were trying to emulate the interior appearance of the great cathedrals and churches of the time. In any case this marked the beginning of the post-and-beam era of construction.

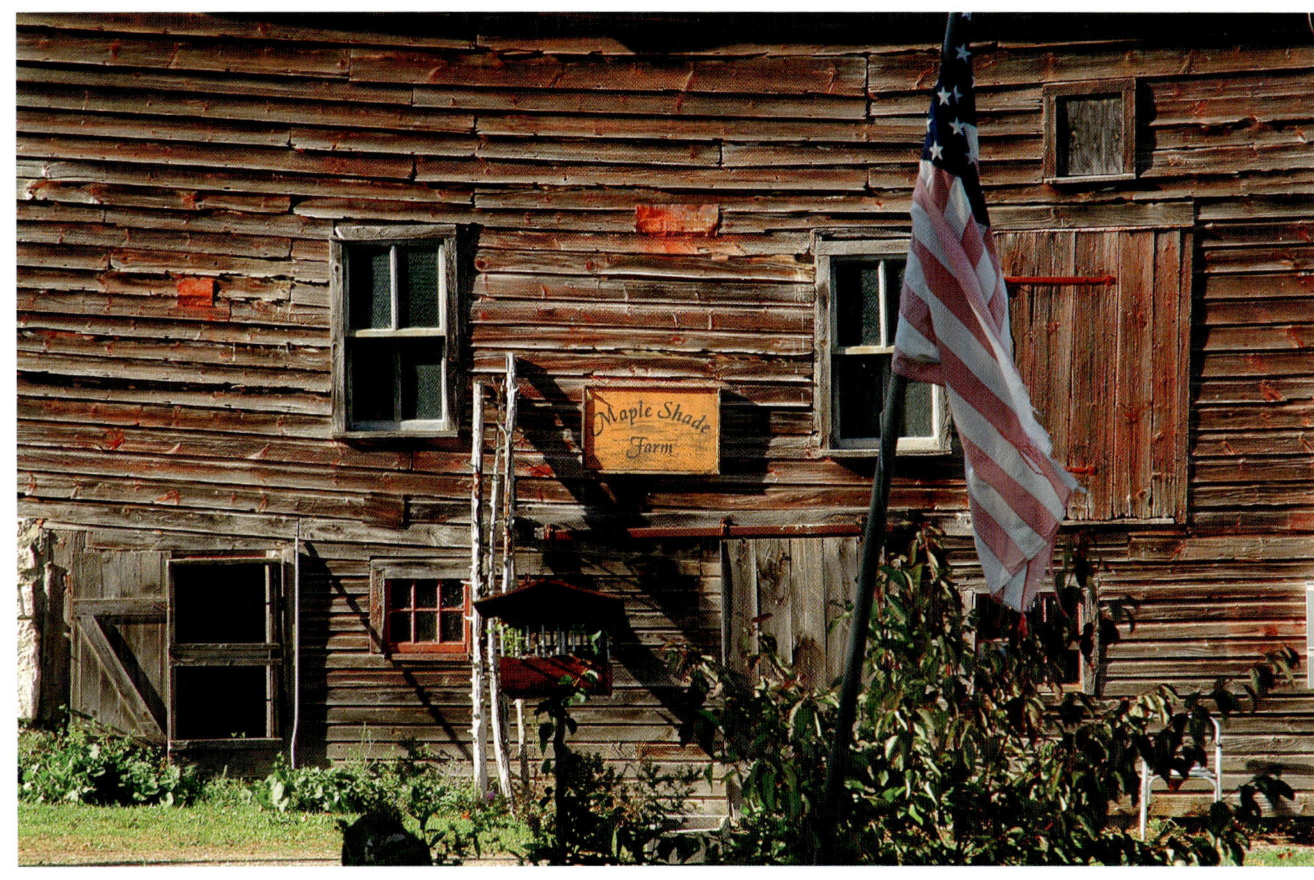

The "Maple Shade Farm" building, which, according to owner Barbara Delmolino, dates to around 1800, was probably a barn first but was converted to a horse stable below and coachmen's quarters above after the property became an inn on the stagecoach routes through the Berkshires.

The first post-and-beam barns that represent the ancestors of American barns began to sprout in England and Germany in the early 1600s. The significant development at this time was that the cruck system was replaced by the more straightforward post-and-beam method that incorporates rectangular post and beam sections, called bents, that are strengthened with braces, and topped by a roof system of rafters and a ridge joist and, often, king or queen posts[3].

It was during this time, of course, that systematic European settlement in North America started to accelerate, first with the Virginia Colony and then, by something of a twist of fate caused by winter gales that blew the Mayflower off course, in Massachusetts and New England. It took almost another 100 years for the first settlers to arrive in the Housatonic River Valley region that was to become Berkshire County.

In those first 100 years or so of this nation's existence the barn was, arguably, the most important building that the early settlers would construct. This is borne out by the fact that most early settlers built their barns, and often inhabited them for a time, before they even thought about building houses. Even if people were not, either by nature, training, or choice, farmers, they had to know how to farm to provide the essential food needs of their families. And so, by necessity, they also had to acquire the skills of barn building.

In the Author's Notes to his book *An Age of Barns*, Eric Sloane writes, "Our forefathers had a deep respect for tradition and the accepted way of doing things. It was their complete adherence to rules that enabled them to do many things well. Because each person helped with the construction of his own house, I assume that building knowledge was passed from generation to generation…in the simplest house framing one can see touches in hand-hewn beams that show a knowledge of classic architecture. Ira Allen, brother of Ethan Allen, wrote of Vermont, 'I am really at a loss in the classification of the inhabitants here. They are all farmers, and again every farmer is a mechanic in some way or other, as the inclination leads or necessity requires. The hand that guides the plow most frequently constructs it'[4]".

By the time the first settlements began to sprout in the Berkshires, around 1730, a distinct style of barn construction had become the accepted model. Almost all barns built in the Berkshires in the first 50-60 years of settlement are classified as "roots barns" which refers to the fact that their designs are based on Old World styles and building techniques. As I indicated above, the greatest influences on the development of barn construction in America were German and English, but it would be the specific demographic makeup of the settlers of any given region that would ultimately determine the type of barn that would come to dominate that landscape. In the Berkshires, as in most of New England, the original population was almost entirely of English decent. Therefore, while there were innovations developing in the materials, style and function of barns being built in other parts of the country, the vast majority of the roots barns that were built in Berkshire County from the 1750s until the early nineteenth century conformed to traditional English barn models.

Right:
Jack Sobon's English barn is, at 25 ft. 6 in. by 34 ft., smaller then the prototypical English barn, but it still adheres exactly to the Pythagorean 3-4-5 rule of proportion. Shown here with the center doors open, you can also see the access doors providing entry to the outer bays. Note that the main doors are off-center.

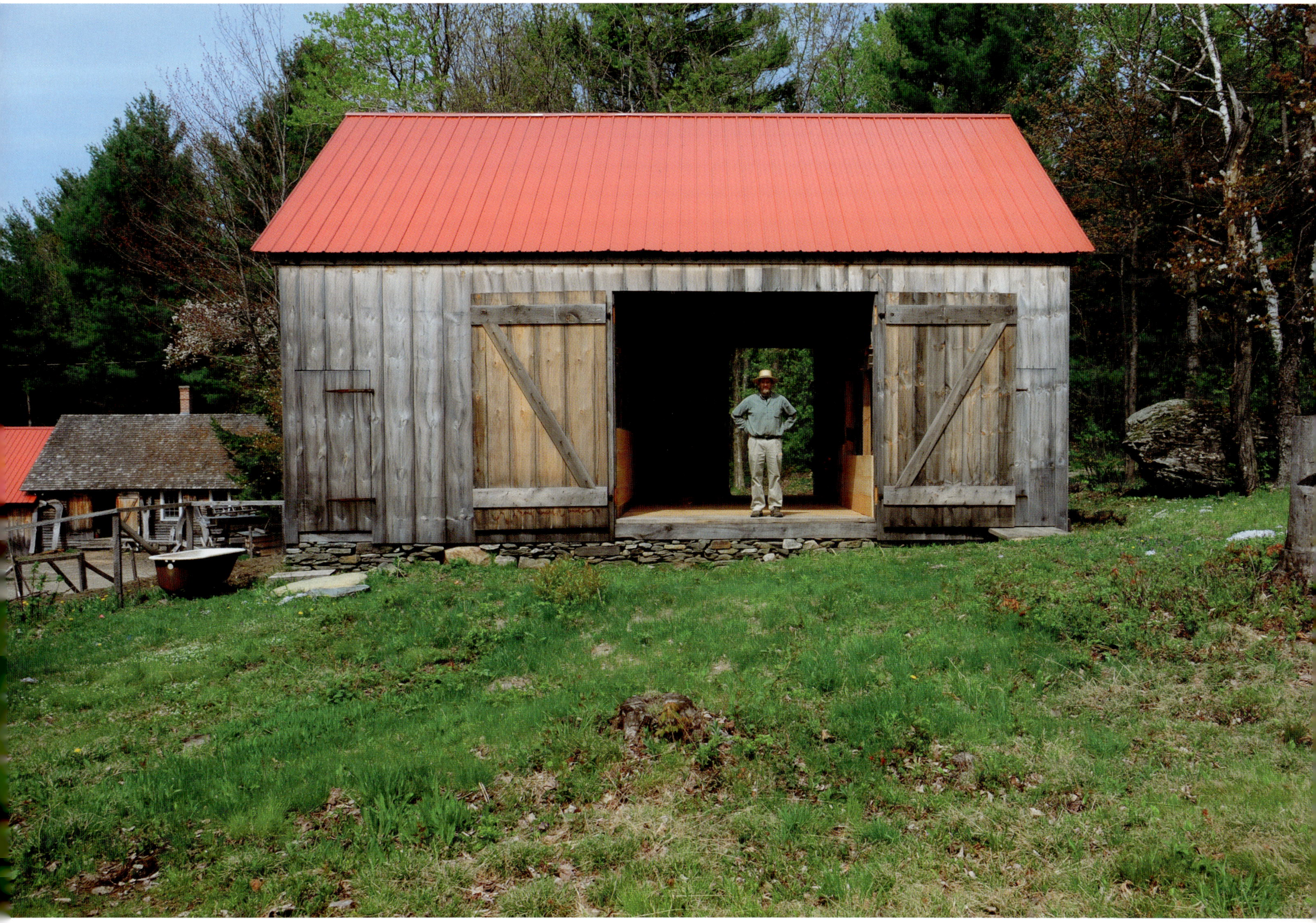

The English Barn

Windsor, Massachusetts, native Jack Sobon is a craftsman builder and a member of the Timber Framers Guild of North America. He apprenticed with Richard Babcock in the 1970s and later branched out on his own. Today he is considered an authority on timber frame construction methods and barn building. His extensive research of the English barn has taken him to England on several occasions and he has been widely published on the subject. Having either built or acted in the role of consultant in the construction of many barn structures in Berkshire County, he decided several years ago to build an authentic reproduction of an 18th century English barn on his property.

Just as Richard Babcock had done, he chose to apply the materials, tools, and methods that had been used by the English settlers. He felled trees on his property, having chosen woods that would be the easiest to work with, while still being sturdy and durable, and hand-hewed many of them to size. Hand-hewing, he says, was less expensive than other options and guaranteed that the barn would be authentic in every way.

The English roots barn that Sobon built is referred to as a three-bay threshing barn, and it is, he says, typical of ninety percent of all the barns built in the first 50 years of settlement in the Berkshires. Three-bay refers to the three interior sections of the barn. The principal function of the barn in both the Old and the New World was to act as a threshing and storage facility for grain—most often wheat, barley, oats, corn and/or hay.

These English roots barns were based on a very specific plan that few settlers deviated from. There were two principal elements to this plan: something called the "scribe rule," and the Pythagorean Theorem.

Pythagorean Theory established the 3-4-5-ratio for the dimensions of a right angle triangle, and it was applied to the primary dimensions of the barn. The most common exterior dimensions of English barns, for example, were 30 ft. x 40ft., and the usual roof pitch was the 9-in-12, an extrapolation of the rule. This allowed for easy confirmation that the building was square, and the roof line even. Builders knew, for example, that their structure was perfectly square when the distance from the end of a 30 ft. wall sill to the end of a 40 ft. wall sill measured 50 ft.

The scribe rule refers to a methodology for piecing together the internal frame of the barn. This was an English method that had been developed to allow builders to work with crooked timber. By the 1500s most of the old growth forests of England had been completely harvested after hundreds of years of development. What tall, straight trees were left had been claimed, and closely guarded, by either the church, for building churches and cathedrals, or by the crown for use in the absolutely essential shipbuilding industry—Britain's merchant fleet was, and would continue to be, the engine of its economy, and the navy was the principal asset for national security. The cruck design discussed above was also attributable to this paradigm.

Utilizing the scribe rule made any form of construction extremely labor-intensive because it required that each section of the frame was pre-assembled on a flat area of ground, marked, disassembled, and then put back together in place. The timber was scarred with reference marks and numbers that would help the builder identify where particular pieces of timber would be joined, and to establish a perpendicular plane for leveling and plumbing the sectional and outer walls, or bents, of the frame. Sobon notes that barn raisings of the popularized sort, involving whole communities of people, were not generally staged here in the Berkshires because, first, employment of this method of construction required that the final assembly of the frame was done piece by piece, and not by completing sections flat and then lifting them into place (which would require substantial manpower) and, second, it was simply not an English tradition.

Fifty years ago, the exact purpose of the reference marks, and the general methodology of building according to the scribe rule, had presented a mystery to people like Richard Babcock. Reference marks, for instance, were only scored into the wood on the layout side, and that meant that they were mostly hidden from sight behind siding on the outside of the barn. Therefore, the marks that were visible, which were almost always on both sides of the center threshing bay, seemed random. Furthermore, according to most authorities on barns, the early builders were either the settlers/farmers, themselves, or craftsman carpenters who closely guarded the secrets and nuances of such things as joinery and assembly, and expected apprentice carpenters to learn the craft through their own keen observation, just as the craftsmen had done. It was, in a sense, an unofficial guild that one gained entry into by proving himself on the job. These craftsmen, already in possession of all of the information and details they needed to successfully put up a barn, didn't require plans or drawings. And so, since very little documentation exists from these pioneering barn builders, much of the knowledge about building methods has evolved through the work of modern-day builders who, in a sense, reverse-engineered the barn. As barns were deconstructed crucial elements, like the reference marks on the outside of the frame, were revealed and it became easier to understand the order and process by which the barn had been erected. Deconstruction also revealed the hidden secrets of mortise and tenon joinery.

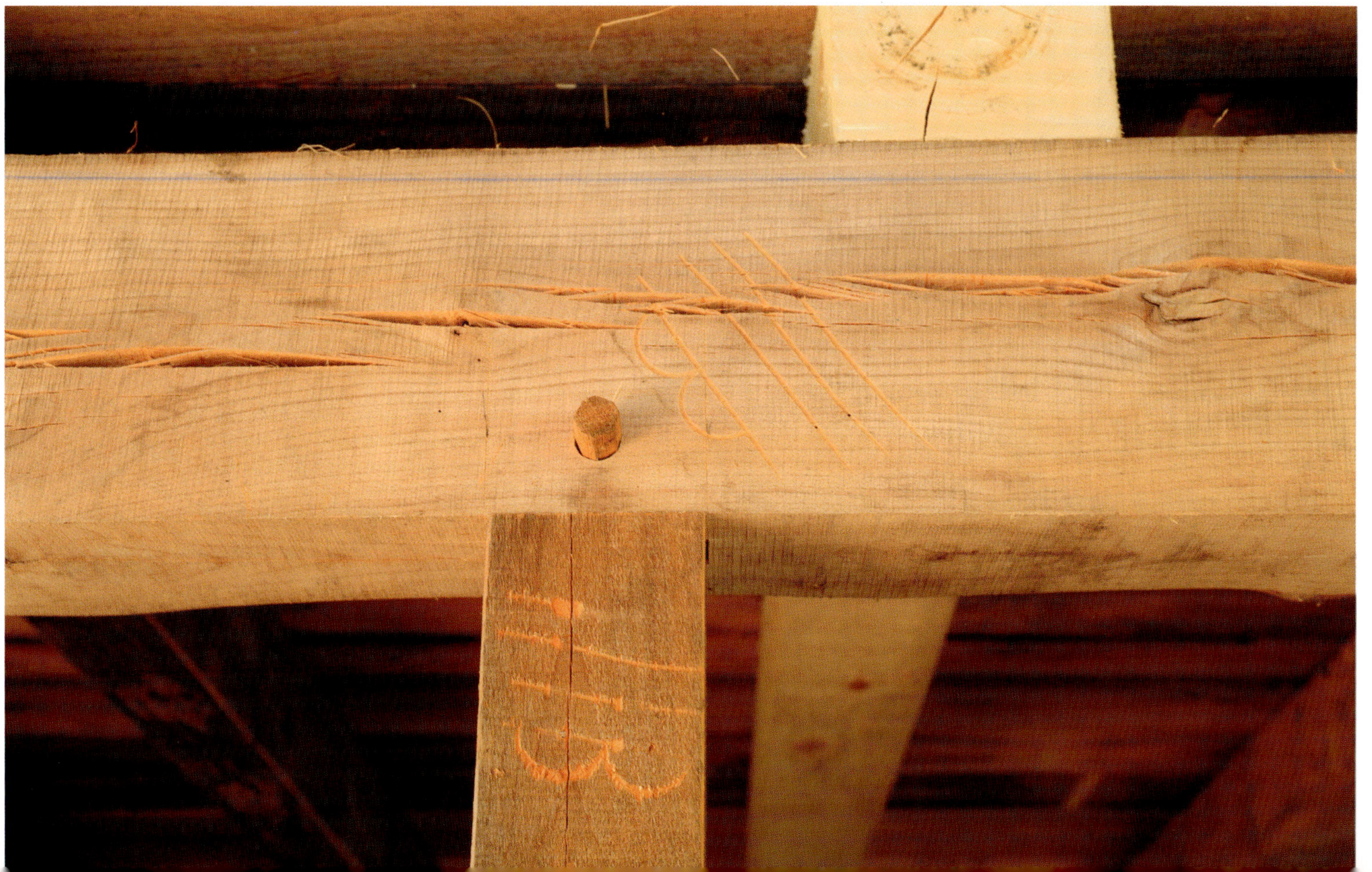

Identical scribe marks appear on a center post and crossbeam of a bent in Sobon's barn. True to form, the marks appear on the side of the bent that faces into the threshing bay.

Beside the scribe marks, there are two other features of the scribe-rule built barn that are giveaways—the jowled, or gunstock, outer posts of the bents, and the accompanying tying joint. The gunstock post refers to a post that tapers out at the top, towards the inside of the barn, over the last quarter to third of its height. By forming it this way the extra width created at the top of the post provided the necessary surface area to accommodate the critical tying joint where the outer post of a bent, the top beam, the wall plate, and a rafter were joined, or tied, together with an interlocking system of mortises and tenons. The tying joint first appeared in structures built in England in the thirteenth century.

The English typically built their barns so the roof joist ran roughly east-west and the principal door opened to the south from one of the long, or eaves, sides. They would contain three internal bays of unequal widths so that the main door was slightly offset from center. Inside, there were two sections of framing, or bents, that create the division of space between the bays. The bays ran the width of the barn.

The simplest version of the three-bay threshing barn was as a grain storage facility. The narrowest of the three bays was most commonly on the east end. In the example of the prototypical 30 x 40 barn, the east bay would be roughly 10 ft. wide. The center bay, with its full-width double doors at both ends, was typically 11 ft. wide, but could range from 10 ft. to 14 ft., allowing the farmer to wheel a fully-laden cart, or wagon, into the barn, fork his grain into one of the side bays, or mows, and then drive straight out the other end. Presumably different types of grain would be stored on different sides. The west bay would measure anywhere from 16ft. to 20ft. wide. The two outer bays usually had dirt floors while the all-important center bay would have a finished floor. The center bay had particular importance because it also served as the grain threshing floor. Once harvested, grain would be carted into the barn and then dried on racks of loose poles set across the top beams of the center bay creating a temporary ceiling about 7 ft. above the bay floor. The dry grain stalks would then be dropped down to the floor and threshed and winnowed there. The center bay floor, therefore, had to be made of hard wood to withstand the constant beating of the threshing sticks (which looked like skinnier, elongated versions of nunchucks) and had to be tightly fitted, with no gapping between boards, to ensure that no grain would fall through during the threshing process. In any given configuration of the English barn, the center bay always served the same purpose.

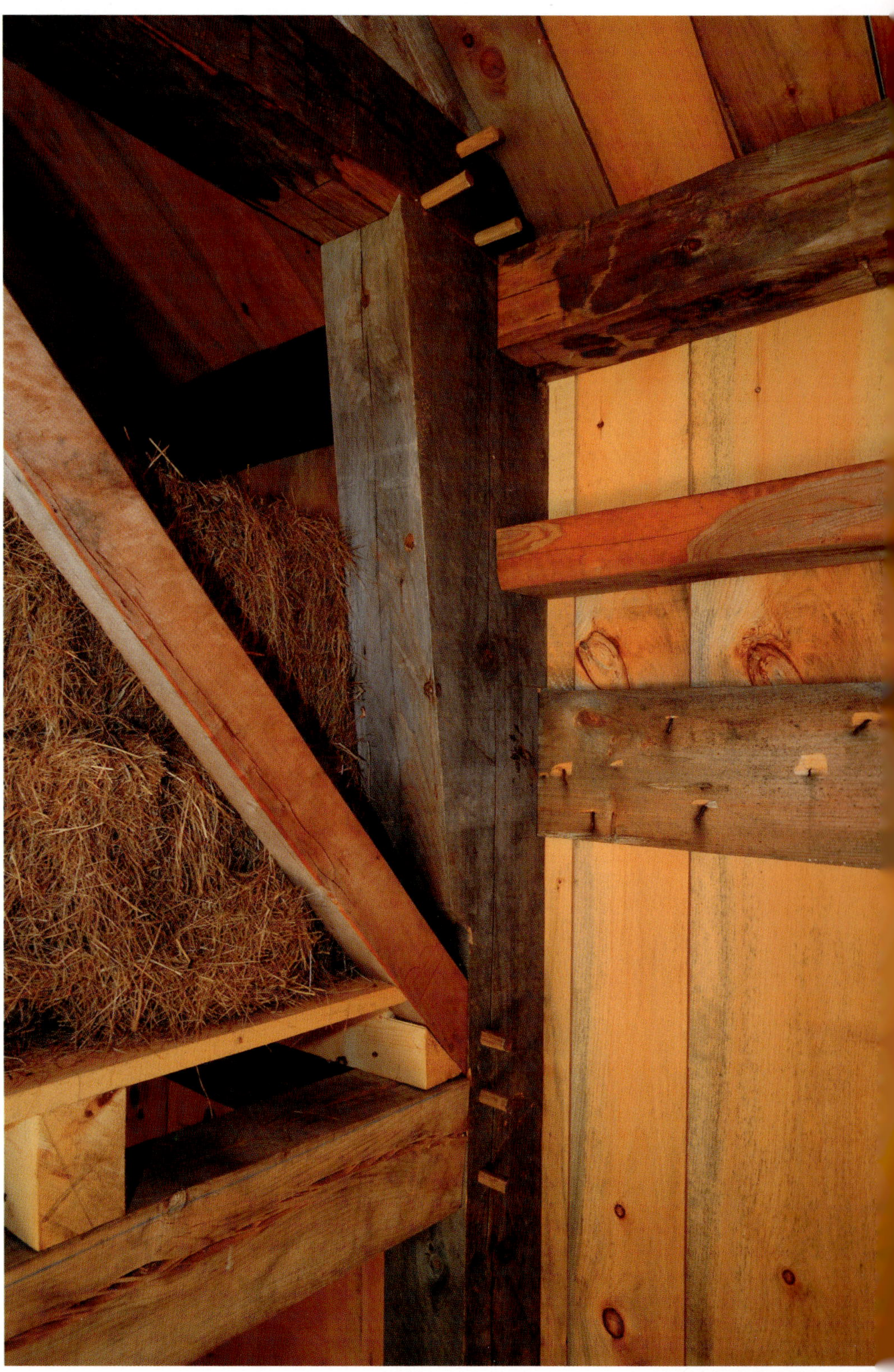

In his article *The English Barn in America* Sobon explains that the framed wooden threshing bay floor was a new innovation. In the Old World the English had traditionally created surfaces out of tile, bricks, flagstones, hard-packed clay and even oak planks laid out over a tamped-down earthen bed[5].

It is interesting to note here that, during our conversations, Jack Sobon also pointed out to me that the word threshold owes its existence to the English roots barn. While threshing and winnowing the raw grain farmers would place a wide board—something akin to a 2x8 or 2x10—on end across the opening of the center bay doorways. This would prevent the separated grain from being blown off the threshing floor, and out of the barn, by any breezes or gusts of wind, while allowing the lighter chaff to fly away. Stepping into the barn while threshing was being done, then, required stepping over this board that held in the threshed grain—the thresh-hold.

While it was typical for the farm layout of Dutch, German, and other ethnic groups that settled to the west and south of New England and the Berkshires to have one large barn structure accommodating many uses, the British had traditionally built several relatively smaller buildings each fulfilling a unique purpose. But the essential utility of the English barn that, with a little ingenuity, the English settlers discovered was that it could act as a multi-functional space. As Sobon's barn illustrates, New World English framers could easily modify the design to accommodate cattle, work animals, and grain all under one roof.

A common alteration would have been to add a large hinged door to the left side of the south wall and divide the front 10 ft. of the west bay into either horse or ox stalls. A floor would be created above this area to preserve as much storage area as possible. In the barns that Sobon and other researchers analyzed they could easily determine the type of animal housed in this area by the height of floor above—6 ft. for oxen, and 7 ft. for horses. The remaining space in the west bay, an area approximately 19 ft. x 20 ft., would remain a hay (or grain) mow from the earth floor to the roof.

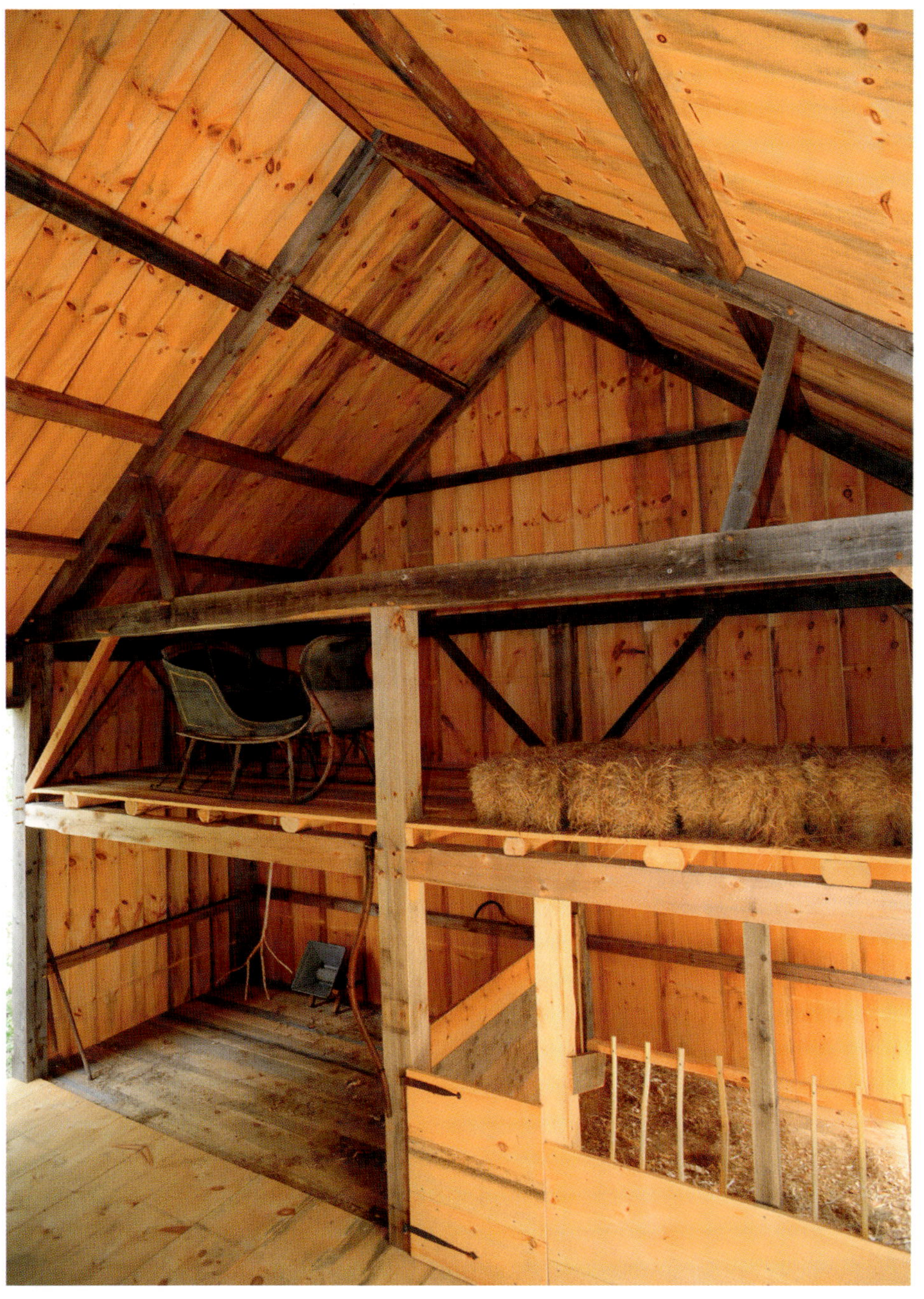

View from the loft area of the west bay of Sobon's barn showing the cow stanchions and the hip-high wall of the east bay. Note that Sobon put a floor in the loft to demonstrate how farmers who converted their basic three-bay barns to multi-functional use retained hay and grain storage space above the cattle stalls. Details of both the bent and roof framing design are also clearly visible.

Left:
This photograph shows a gunstock post with the tying joint that was a hallmark of the scribe rule. The joint is at the top of the post where post, beam, plate, and rafter come together and are secured by an interlocking system of mortise and tenon joints.

The internal design of Sobon's barn is arranged similarly to one that would have been in use on a farm with livestock, work animals, and crops. Externally he added hinged access doors in both corners of the south wall, providing access to the two outer bays from the outside. In the east bay he built stanchions, or stalls, to the left of the outer access doorway, that run half the length of the bay—roughly 12 ft. The stanchions are installed at an angle so that the tops are set back about a foot inside the bent that creates the division between this bay and the center bay. Typically, in the 30 x 40 model of the early settlers, the stanchions would have run the whole length of the bay and accommodated up to 11 cows. Some farmers might have installed flooring above this bay to preserve the maximum available amount of storage space. In the west bay Sobon included stall space for draught animals as described above. On both sides of the center bay where the stalls were located several rows of wide boards would be added to create hip-high dividing walls to contain dirt and animal waste in the stall areas. These were necessary in every English barn to help keep the threshing floor as clean as possible.

On the exterior, most English roots barns employed wide-board vertical siding. This is due to another unique feature of the framing detail. A grooved board channel was chiseled out of the underside of each top plate of the frame along the outside line of the frame posts. This was a simple yet ingenious time, labor, and money saving trick. First, since it ensured that every board of siding could be the same length, each board only had to be cut to length once. This usually took place in the field where the trees were felled. The stacks of siding could then be brought in and installed right off a cart. The groove held each board in place at the top so that they did not have to be nailed there. The only nailing that was required could be done standing at ground level—no scaffolding or ladders were necessary. From the raw material stage to the last nail the process was highly efficient, and this in turn saved time and money. Sobon allowed his siding to dry and cure before attaching it, creating a uniform, tight-fitting look. On original barns, however, siding was often nailed on when it was still wet, or green. As it dried it shrank creating the familiar appearance of many an old barn on which gaps appear regularly between the siding boards. This did have the advantages of allowing some light to enter the building, and providing ventilation[6].

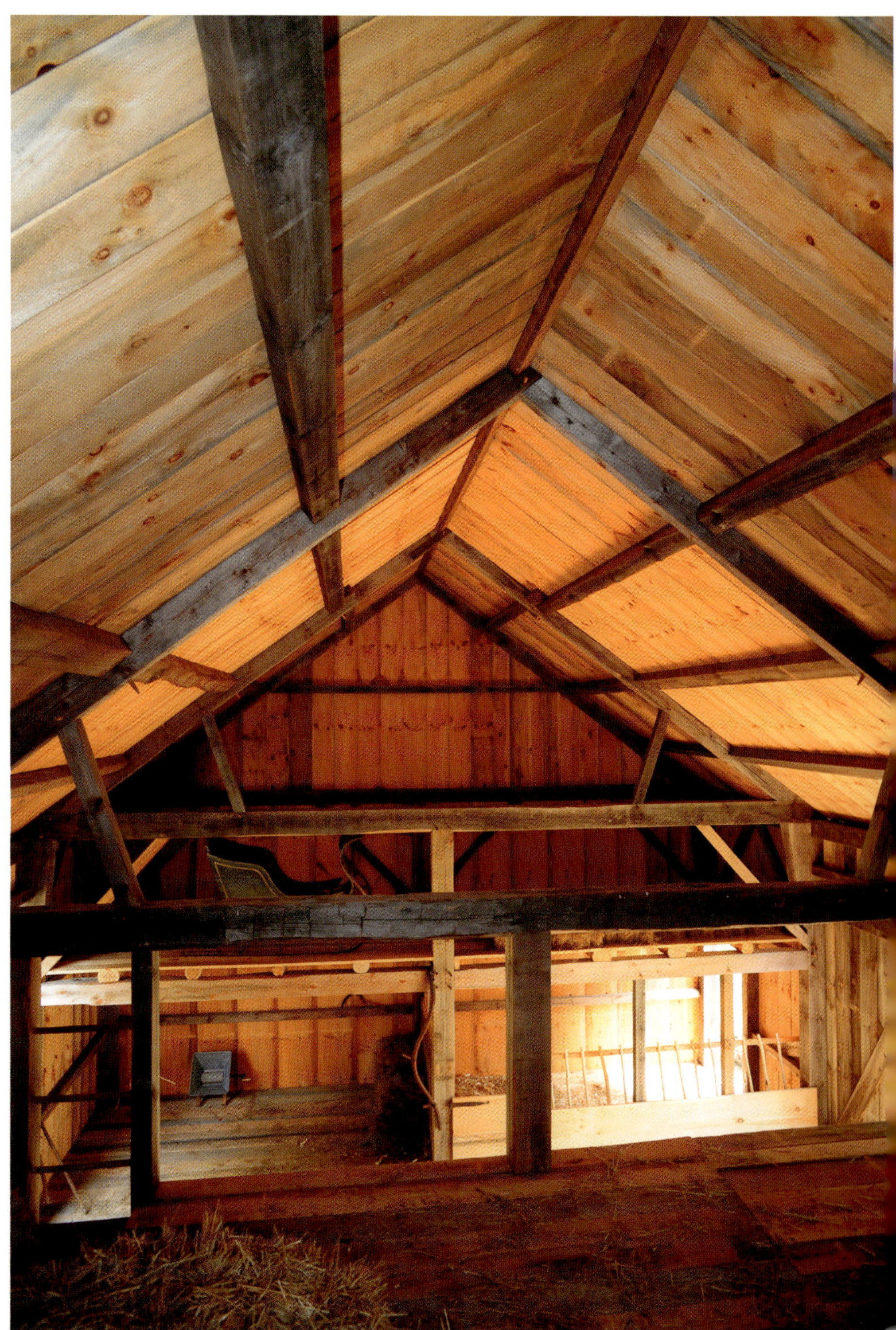

This view from deeper inside the west bay loft area shows three of the four rafter sets, mortised for purlins, that support the entire roof. The overall design of the three-bay barn is eminently simple and efficient in its use of materials.

Sobon constructed the frame of his roof with just four sets of rafters with one set capping each bent. The two internal sets of rafters were strengthened by angle braces, or struts, coming off of the top crossbeam of the bents and angling back toward the rafter, while he added collar ties for rigidity to the two outer sets. The rafters were mortised to accommodate three rows of wide purlins and a ridge pole to which the roof boards were nailed. This design allowed for an unobstructed loft space that one can stand up in and walk through. You will also notice on the exterior photographs that the roof slightly overhangs the outer walls allowing for optimal shedding of rain, snow and other weather-related phenomena. Various framing design drawings I researched show that the roof design varied and may have employed additional rafters and other forms of bracing to add rigidity to the rafters and prevent sagging.

In contrast to the Dutch and Germanic craftsmen in other parts of early colonial America, the builders of English barns throughout New England put issues of function above matters of fashion. For the first 150 years, from the 1650s to 1800, English barns, including those in Berkshire County, had very little decorative embellishment added to their exteriors. At most a farmer might have added patterned swallow holes to one or both gable ends. But even this served a primarily practical purpose as the swallows performed the important function of keeping the barn interior relatively free of insects.

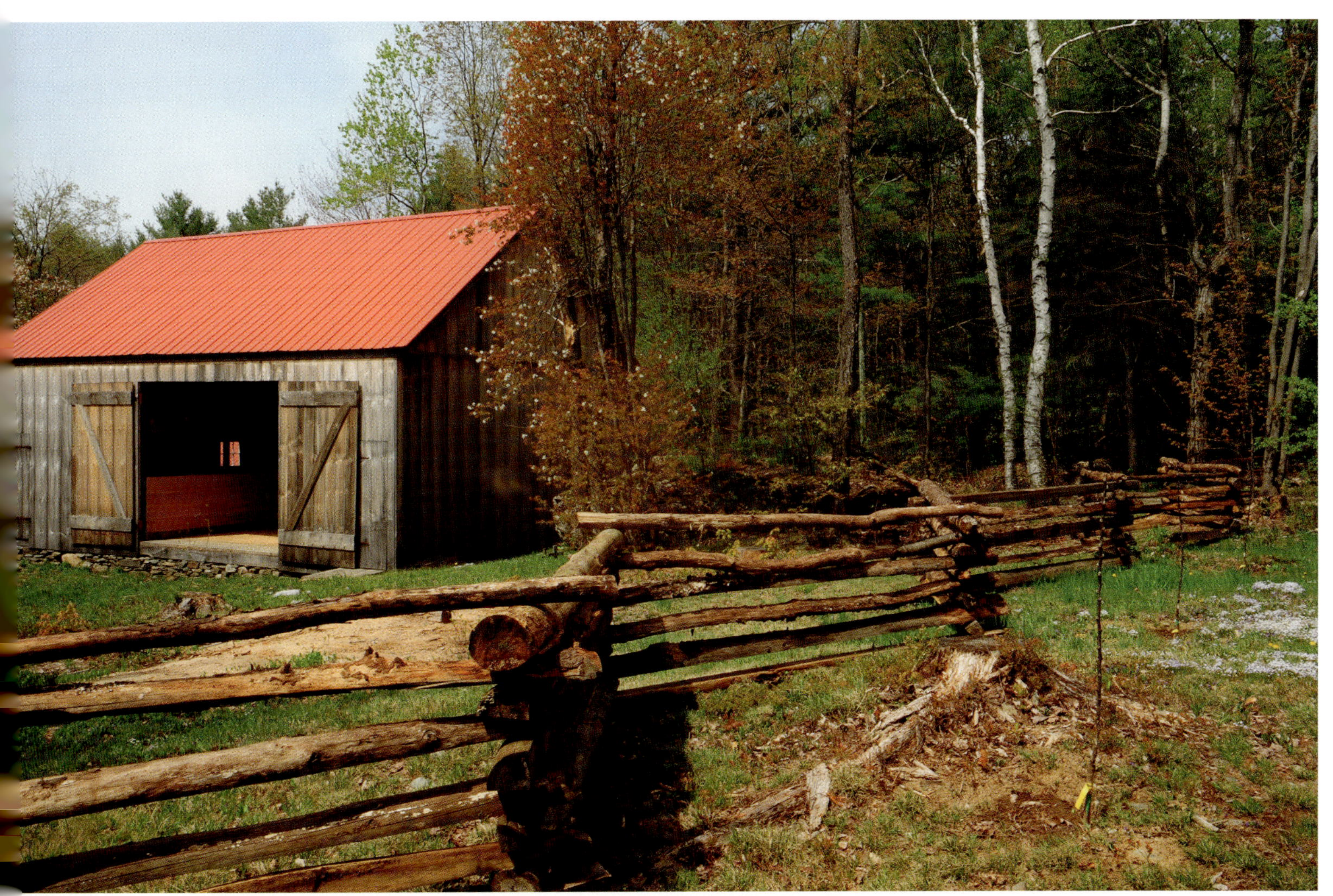

Southeasterly view of Sobon's barn.

Sobon did grant himself some latitude in a few aspects of his barn. The external metal roof sheathing, of course, is not historically accurate. The original builders would have chosen between thatch, shingle, or lapping boards depending on how the roof was framed. Sobon also built a complete dry stone foundation wall that would not, mostly due to cost, have been an option for most of the early settler-farmers. More likely they would have only built up some form of stone foundation pilings to handle the weight carried by the frame posts. But in the functional design and the construction process, Sobon's barn is an excellent representation of what was most commonly observed in the Berkshires landscape of the late 1700s.

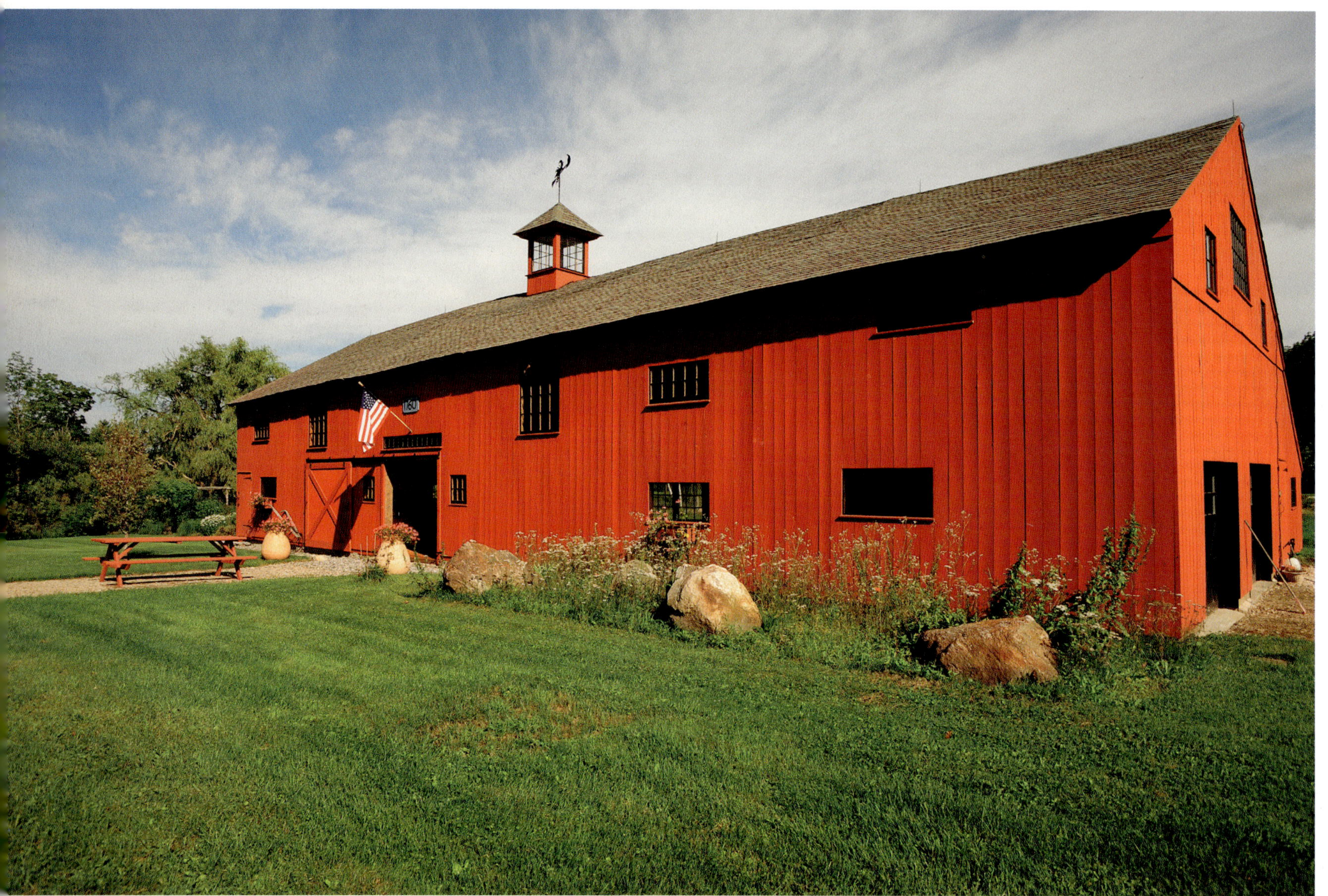

This Sandisfield barn, which is believed to date to 1760, is one of the oldest surviving barns in Berkshire County. It is still standing in its original location and has been reconstituted for mixed use as an entertainment area, a painting studio, a workshop, and storage area. The cupola is not an original feature and probably dates to the 20th century.

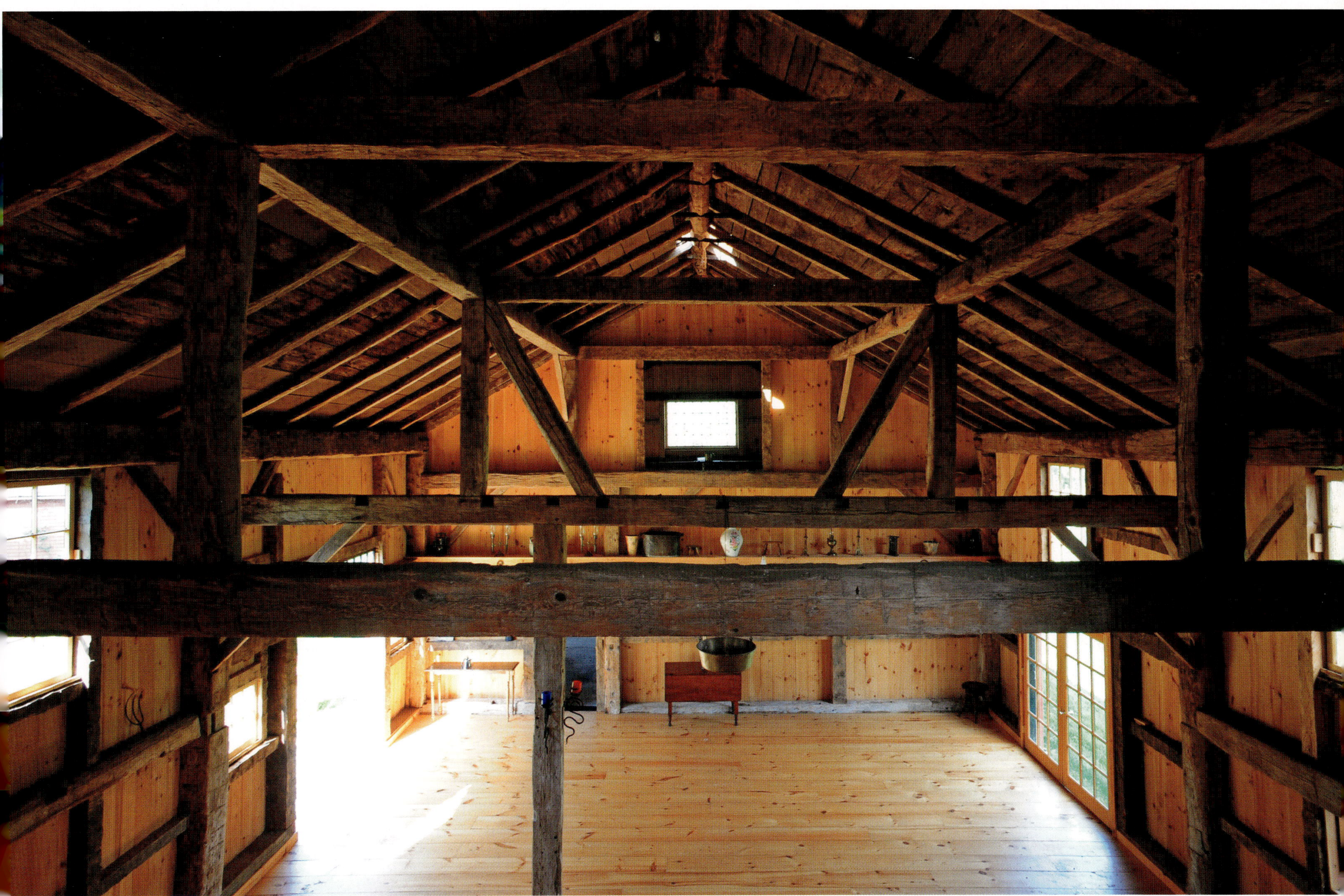

The roof in the Sandisfield barn is supported by a set of queen posts which top off each bent. The original side-entry barn appears to have additions on both ends.

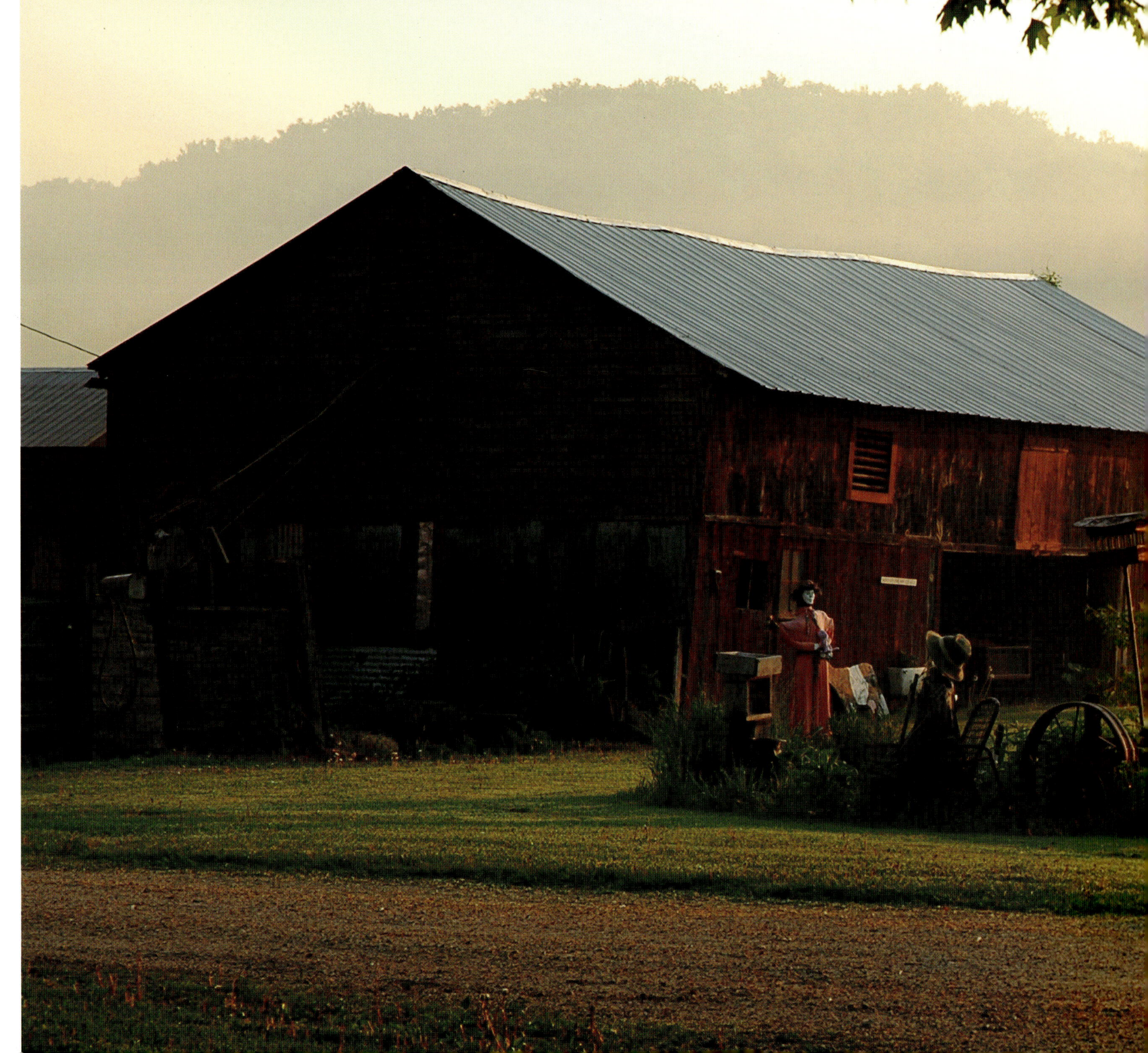

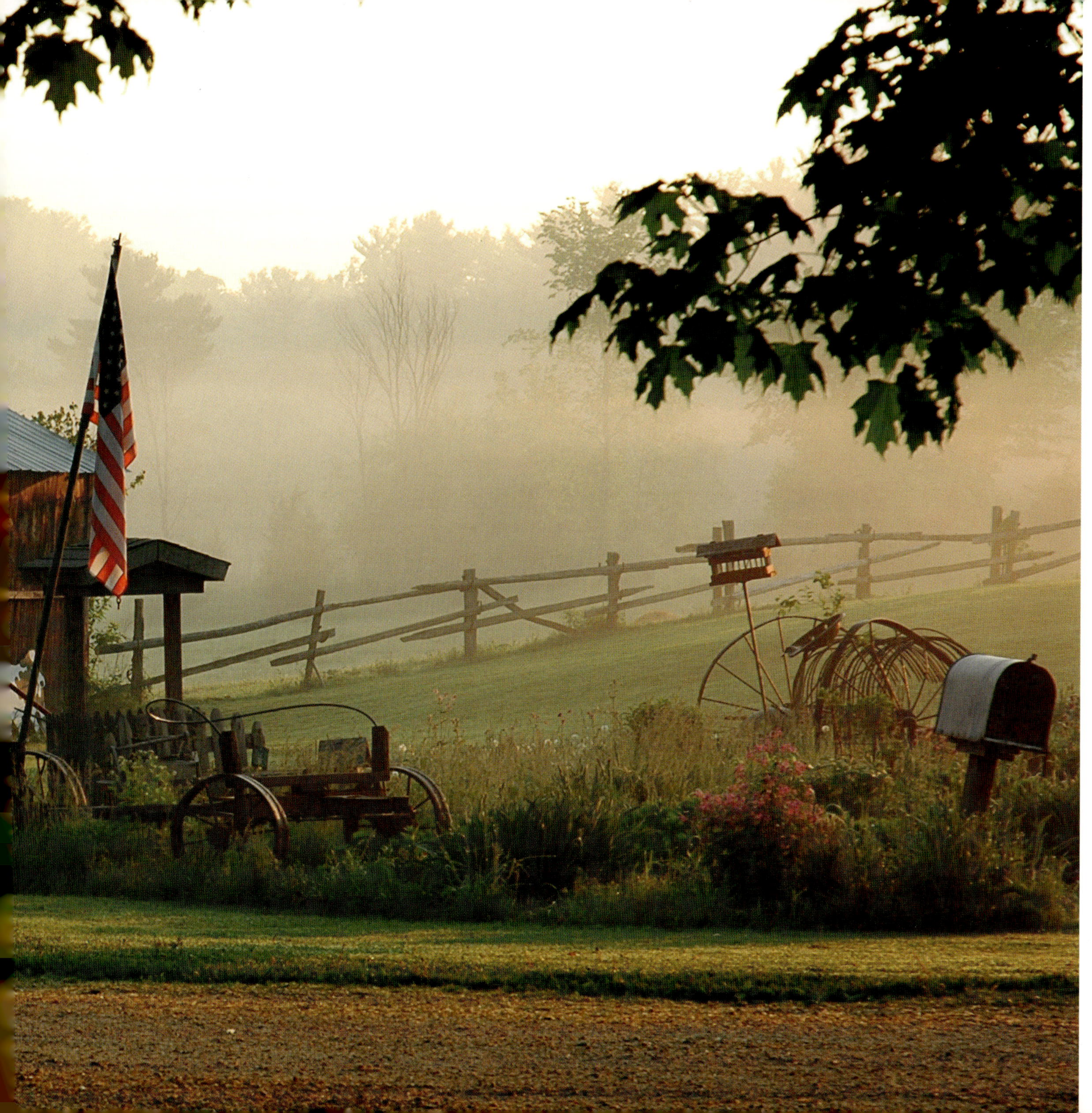

The pole barn at Delmolino's welcomes the first rays of sunrise on a cool spring morning. The Delmolino farm is set in a shallow valley with stunning views of Mt. Everett State Reservation. It sits in the middle of a triangle of roads which makes it one of the most accessible properties in the Berkshires to photograph from all sides, and at all times of day.

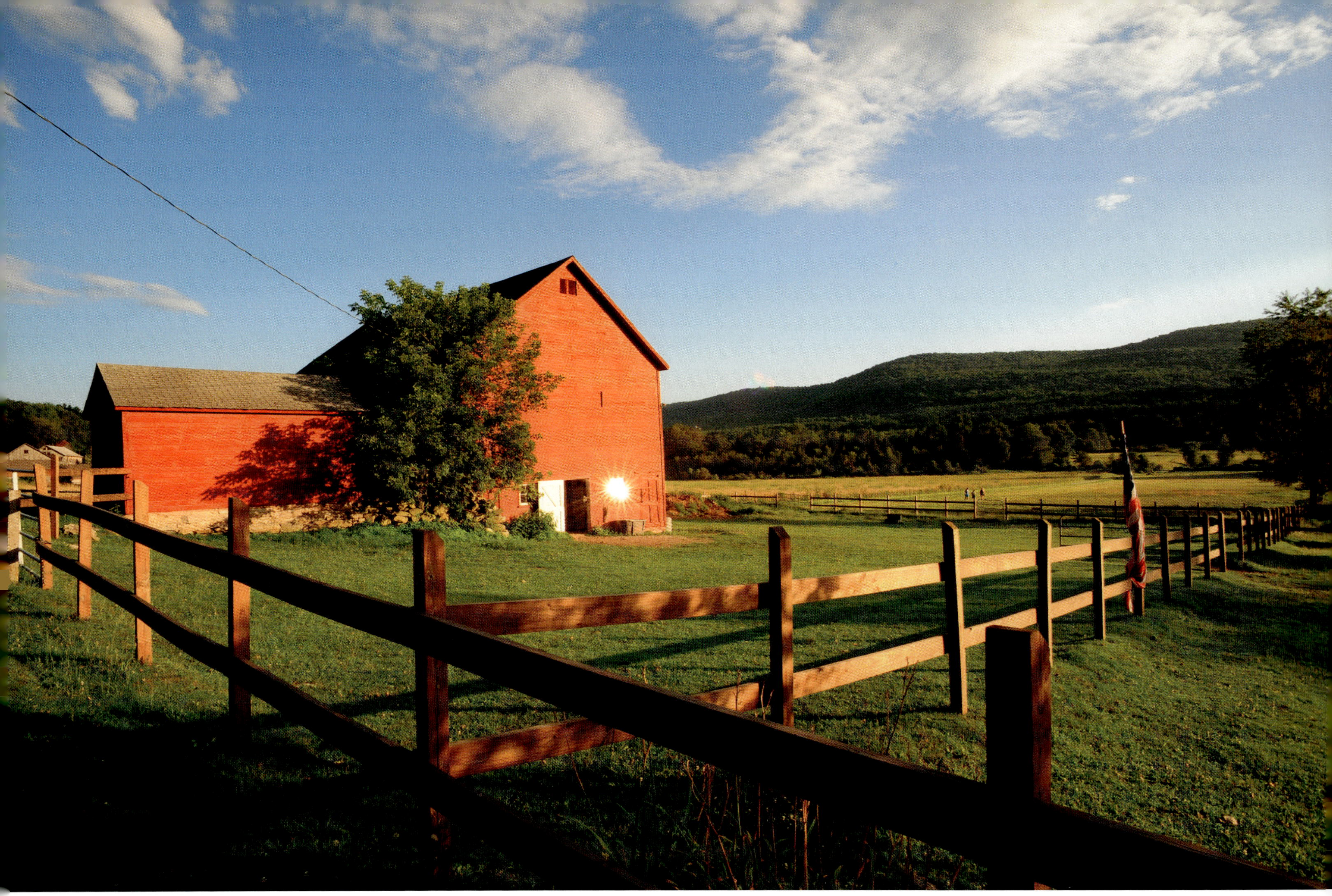

The striking red barn on Patricia Clark's property on Main Street in Tyringham is one of many that was transplanted from the Fernside Shaker Village that is located on the mountainside south of town. The barn probably dates to the mid-1800s.

Evolution of Barns in Berkshire County

Some of the first innovations to barn building in the Berkshires and New England affected the core construction methodology. They were due to a critical new reality the territory presented, which third and fourth generation settlers embraced enthusiastically. As discussed above, the previous 500-1000 years of human habitation, development, exploration, and war in western and central Europe and the British Isles had left much of the continent devoid of virgin forests stocked with tall, straight lumber. The New World, on the other hand, had an abundance of old-growth forests, the likes of which most new arrivals had never seen. Eventually this ushered in the application of the square rule method of construction.

Since town records often make mention of the emergence of the first square rule building, we know that it became general practice around 1800. Many consider this the first truly American innovation in barn building in New England. In a few critical ways the square-rule revolutionized the whole process of barn building. Having long, straight timbers with which to work allowed builders to standardize the individual components of the frame and make them interchangeable. Since all similar parts of the frame—post, brace, beam, plate, etc.—were cut to the same exact specification, parts no longer had to be individually marked or scribed and mapped out for their specific place in the frame. This, in turn, allowed the builder to cut and form any one piece at any time during the process if needed. Thus, it was no longer necessary to pre-assemble and then break up the bents on the ground before final assembly. The square rule method also did away with the need for the gunstock post described above, which makes it very easy to differentiate between scribe rule and square rule barns at a glance. The net effects were that this new method required half the labor and a fraction of the manpower that had been necessary for scribe rule barns. What used to take 6-12 carpenters and apprentices to accomplish now required just one carpenter. Thomas Visser notes that it took just 20-25 years, or by roughly 1820, for the square rule to gain almost universal acceptance among builders[7].

It seems fair and logical to ask why it took nearly 100 years for builders to "discover" this new construction method. I could not find any research material that directly addressed the question so I have had to rely on my own intuition in proposing an answer. The general history of the early English settlers who eventually made it to the Berkshires shows that, while the New World presented limitless opportunity for them, it also posed many novel and harsh realities, not the least of which pertained to weather and the radically different cycle of the annual seasons. In order to survive, they had to work efficiently and expediently, and this meant doing what they already knew how to do best. There was no time to experiment with news ways of building barns, and utilizing the scribe-rule method did not produce an inferior or compromised barn, so why change something that wasn't broken or flawed.

We do know, however, that newer technology and standardization of materials also hastened the acceptance of the square rule. Better saws, beginning with water-powered sash saws, meant that it was no longer necessary, or cheaper, to hand-hew timbers. Wood could now be cut to uniform dimensions allowing for more precise and predictable fitting of the parts. Through most of the 1700s roofing and siding were attached with individually hand-wrought nails made by blacksmiths. Common types of nails were rosehead, butterfly, and L-heads. By the turn of the century machines had been developed to churn out standardized nails beginning with the Type A[8]. The introduction of mass produced nails might seem like a minor advancement in the greater scheme of things, but in truth it was an extremely valuable leap forward that both reduced construction time and associated labor costs, and accelerated the movement to smaller, easier to manage sizes of timber for all phases and types of construction.

Early modifications to the outer appearance of the barn were generally subtle. Some of these went hand-in-hand or created cause-and-effect alterations to barn design. In many cases the research does not solve the question of which change came first, but the correlations can clearly be traced in many excerpted writings and farming bulletins of the time. One example is the transom window. Initially this was as simple as a thin, unsided opening, probably no more than ten inches to a foot high, above the whole width of the main barn door. Early versions would have been covered by a sort of horizontal shutter-board, hinged at the top, and held open by hook so that it could be opened during daylight hours. Clearly this was the first attempt to bring more light into the barn. It wasn't long before the transom window was finished off with glass, as is evidenced in articles written in the early 1820s[9].

Right:
Another beautiful barn in Tyringham is this side-entry English bank, or side-hill, barn with its decorative cupola and unique spiraling gable window. The accompanying barn, built at a right angle to it, helps create a courtyard at the back of the farmhouse. I was unable to obtain dates for the construction of these barns. Many side-hill barns similar to this one, however, were built between 1820 and 1860 throughout New England.

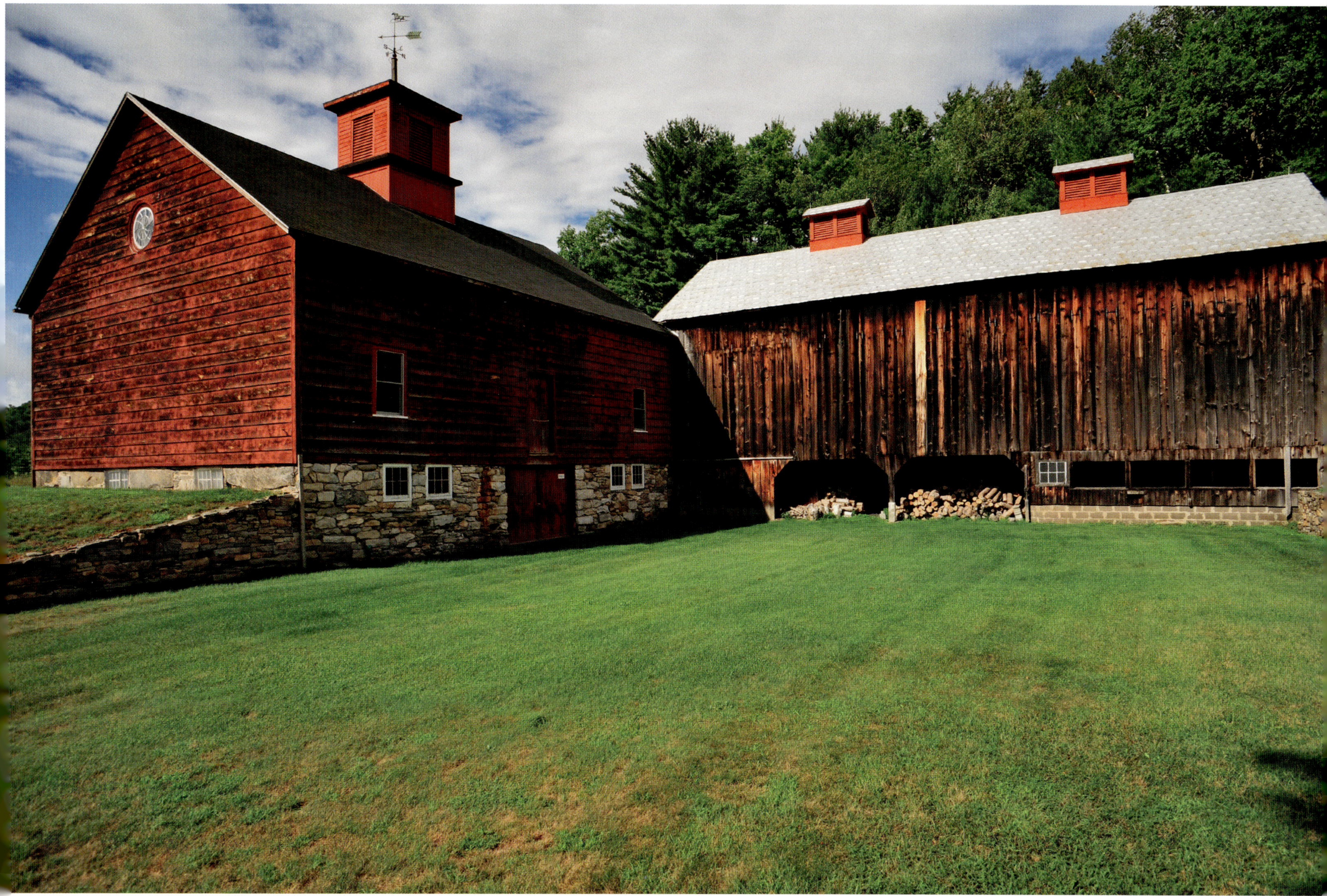

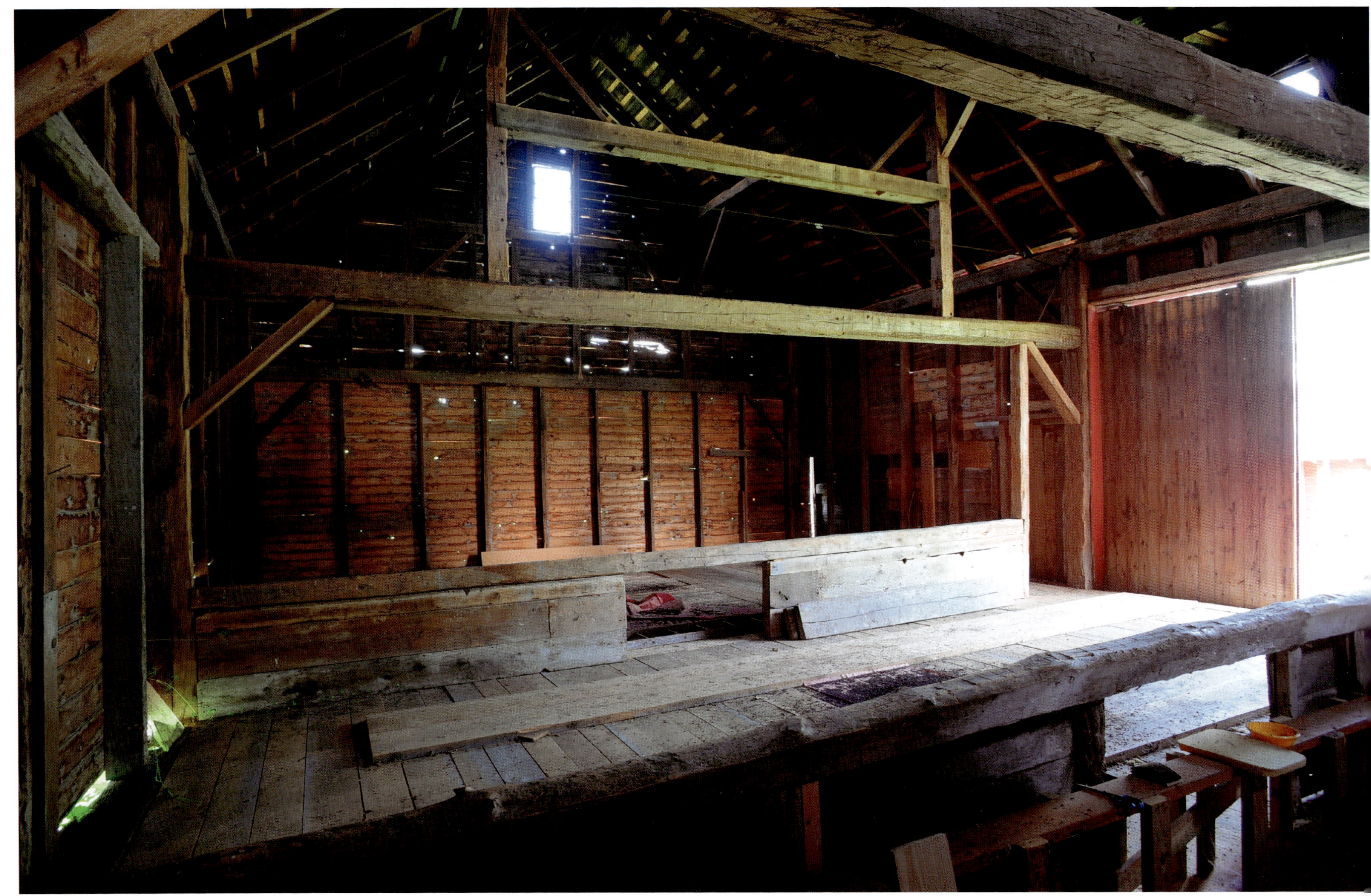

The interior of this side entry barn suggests that it is an English barn built in the early-to-mid 1800s. Much of the timber has been replaced which makes it difficult to determine its age. The studded end wall also defies any possibility of earlier construction, although it could have been retrofitted at some point to accommodate a new style of siding.

Many of the early modifications to the English barn design could appear to be purely aesthetic since they are mostly visible from the outside. The truth is, however, that almost every one of them was developed to enhance life for the livestock that occupied the barns.

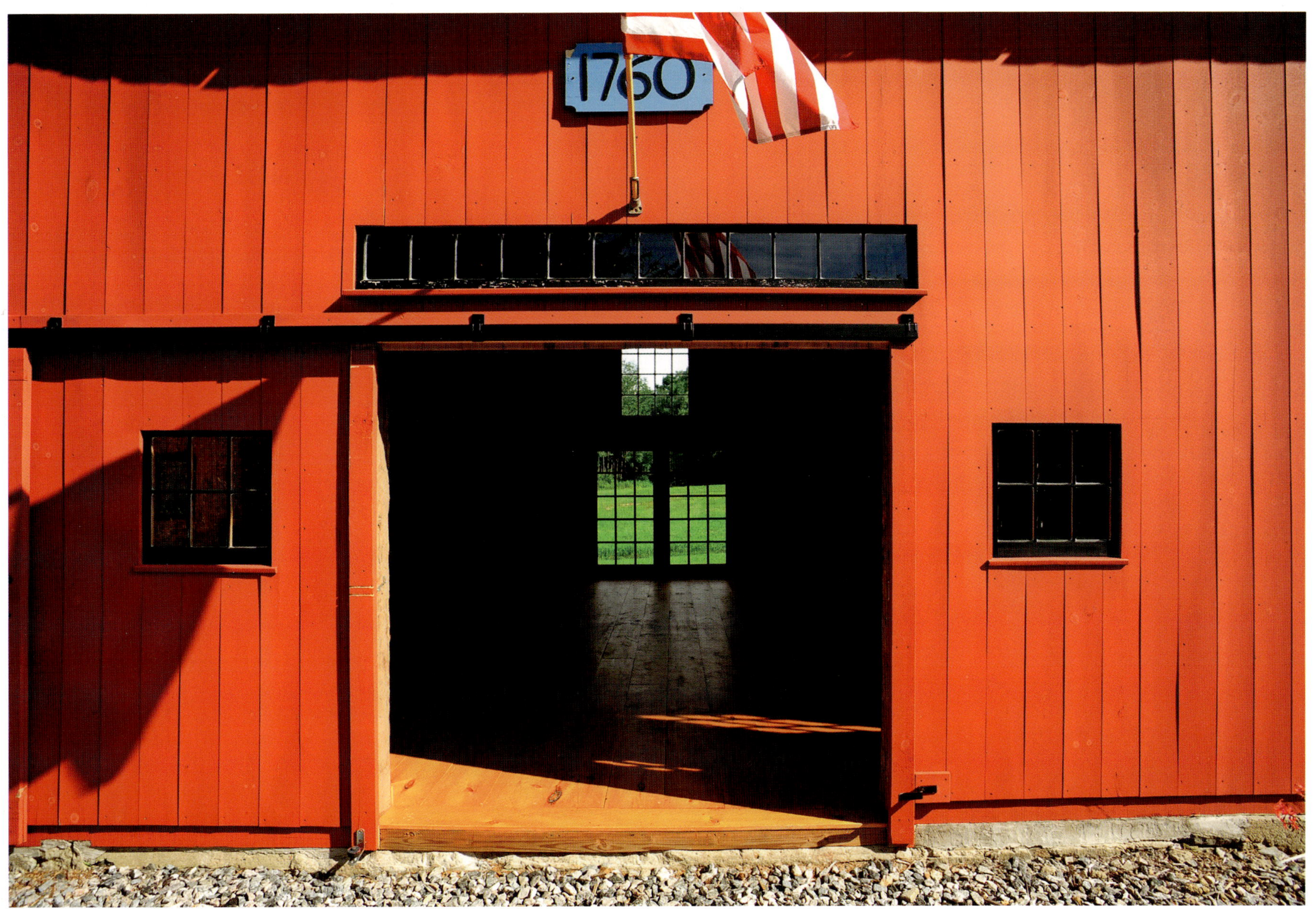

An example of a transom window above the entry to the Sandisfield barn. The window could originally have been an open space with a hinged shutter-board. The glazed glass panels may have been added later.

The three-bay roots barn came across the ocean with the first settlers and it served them well in the first 100 years of colonization. In England the barn had been perfectly suited to both the climate and the nature of farming where grains and crops accounted for most of the use of the barn and animals could spend most of the year outside. In the New World, however, the harsh seasonal variations in New England required animals to be sheltered for up to five months. And, as animal husbandry and dairy farming became more prevalent, it became clear that the barn had to evolve. Healthy animals needed sunlight. And so, as described above, transom windows became more standard. Soon glazed side windows, aligned with the cow stalls, began to appear. Often farmers recycled windows from homes that were torn down or remodeled. Double hung, mullioned windows, for example, were a popular choice for recycled use on barns[10].

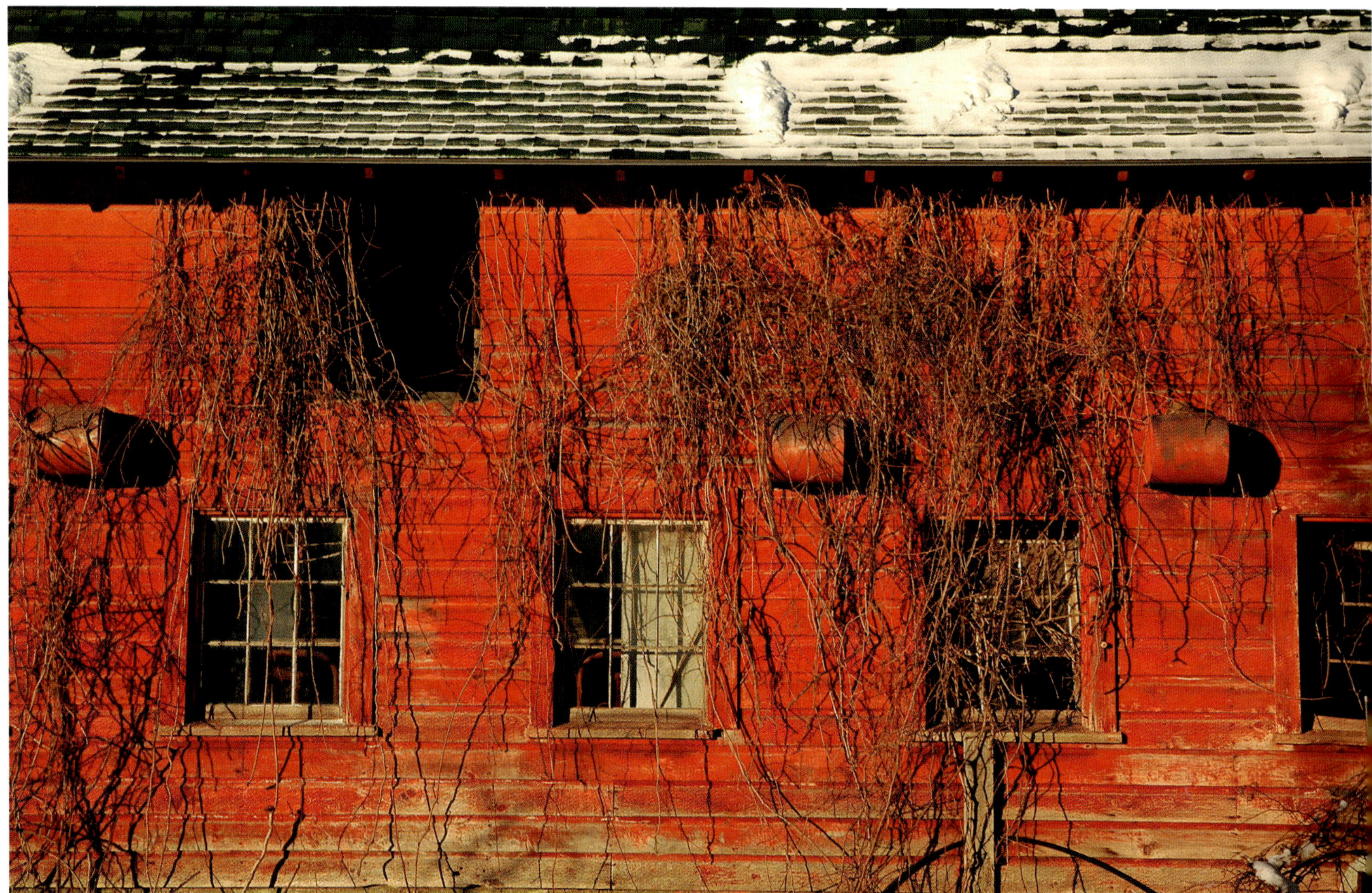

An example of side windows on a dairy barn. No date of construction for this derelict barn was available.

Severe winter weather was the reason for another related series of changes. By the early 1800s farmers had come to realize how weather, especially the cold of winter, affected their animals. When cattle were cold, for example, they needed more food, and eating more food, in turn, reduced their milk production. Obviously this was a less-than-desirable arrangement. Original barns covered with siding boards that had shrunk and gapped were drafty and very cold in the winter, resulting in these changes in the cows' habits. Soon farmers began tinkering with solutions. The quickest, simplest, and least expensive fix was to nail additional thin boards over the gaps from the inside. This is known as double boarding. It was a perfect solution for addressing the problem of cold temperatures inside the barn since it used little new material and didn't require re-siding the entire barn. It was immediately apparent, however, that this made the barn almost pitch black inside when the doors were closed. Obviously, during the nearly five months of winter, the doors had to be kept closed almost all of the time to maintain warmth in the barn. This, then, cycles back to the issue of light.

So you can see how these are, if not cause-and-effect, at least inseparable and interrelated issues that brought about numerous changes to the barns at roughly the same time. My research does not solve the question of which changes preceded which, but it clearly shows that they were all beginning to appear around of the turn of the nineteenth century.

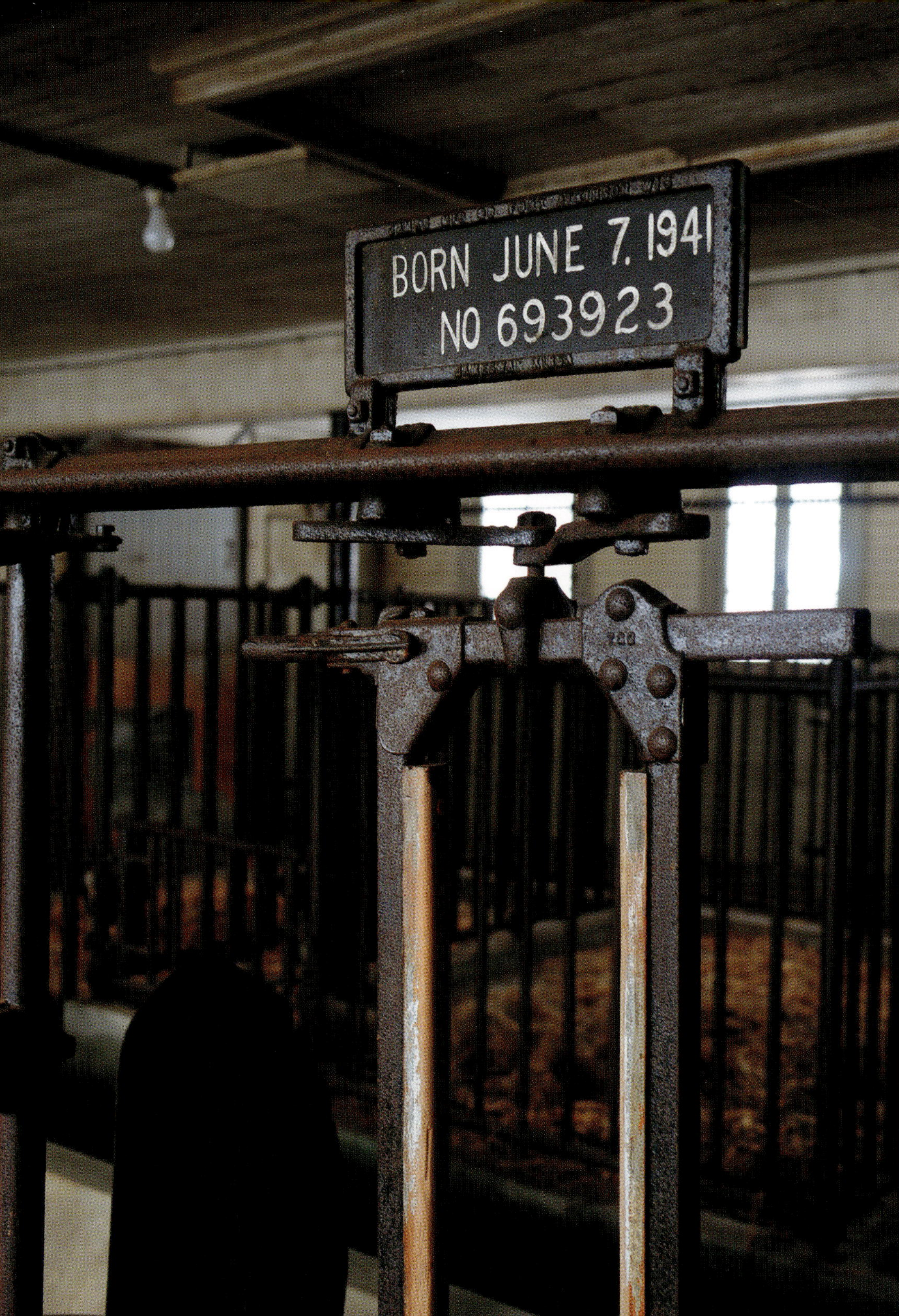

Light streams into the cow stalls in the barn at Naumkeag, Stockbridge, where the loving owners created name plates, including date of birth, for each cow.

Furthermore, the chain of cause-and-effect does not end here. As farmers sought to improve conditions for their cattle and other animals by better managing barn temperature and bringing in more sunlight, they were sealing off the barn and reducing ventilation and air circulation. By the 1820s and 1830s many farmers had also begun to build cellars with their barns and these were often used for storing manure. The net effect of these changes was that barns were now found to be musty and excessively humid which promoted rot in the barn structure and disease among the livestock. The challenge now was to invent new ways to adequately ventilate the barn. Thus, by the 1840s cupolas, which were essentially ventilator shafts with louvered roof boxes, began appearing on top of barns. By the mid-1850s gable vents and windows were also being added to improve air and light quality. Soon cupolas became elaborate design features that served an essential function. Often they were topped off with ornate weather vanes and the box housings were given stylish finishing details. Visser makes mention of Italianate, Arabesque, Gothic and even Victorian-era Queen Anne style cupolas[11]. Eventually the Louden Machinery Company in Indiana began producing the rounded steel ventilators that are still visible on barns throughout New England and the Berkshires today[12].

The dairy barn at Ziemba's farm in Adams has a simple venting cupola that would have been typical of early versions added to barns for practical purposes.

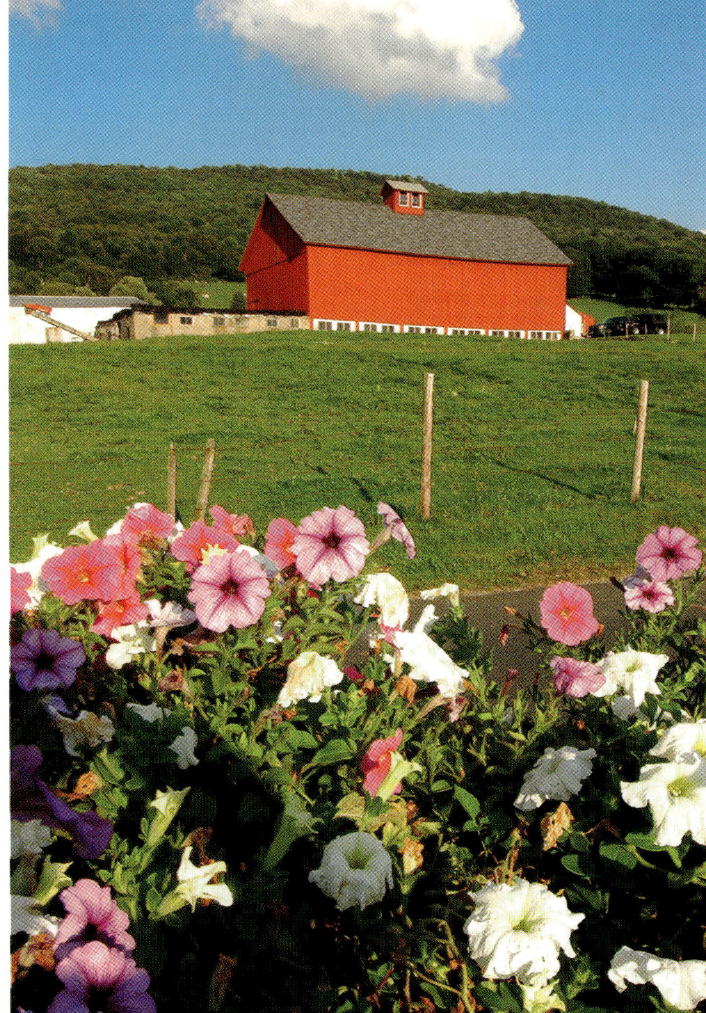

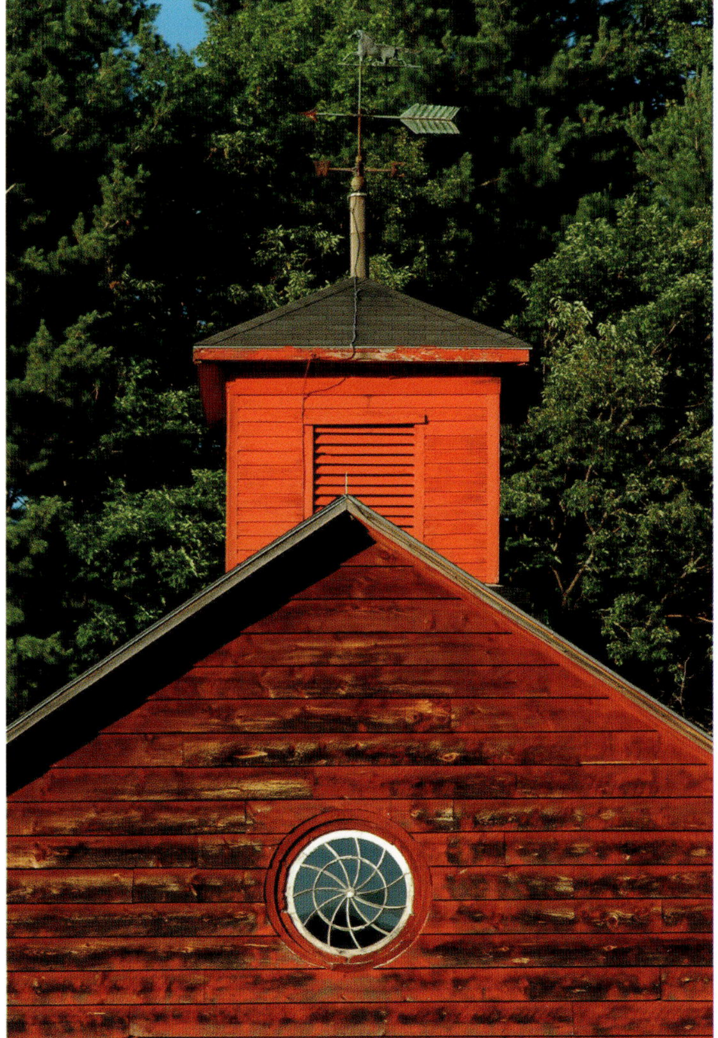

The cupola on the Tyringham side-hill barn is an example of how builders began to play with notions of decoration. The spiral gable window is clearly visible here. This is the only example of a window like this that I have observed.

By the mid-1800s there were many other simple and, sometimes, purely cosmetic alterations made to the appearance of barns. Different types of siding, like cedar shingle and board and batten, a variation of the earlier solution to gapping boards, came into fashion. From 1860-1920 red-painted clapboard with white trim became popular and around 1920 novelty tongue-and-groove siding was introduced. All of these siding styles have endured to the present time.

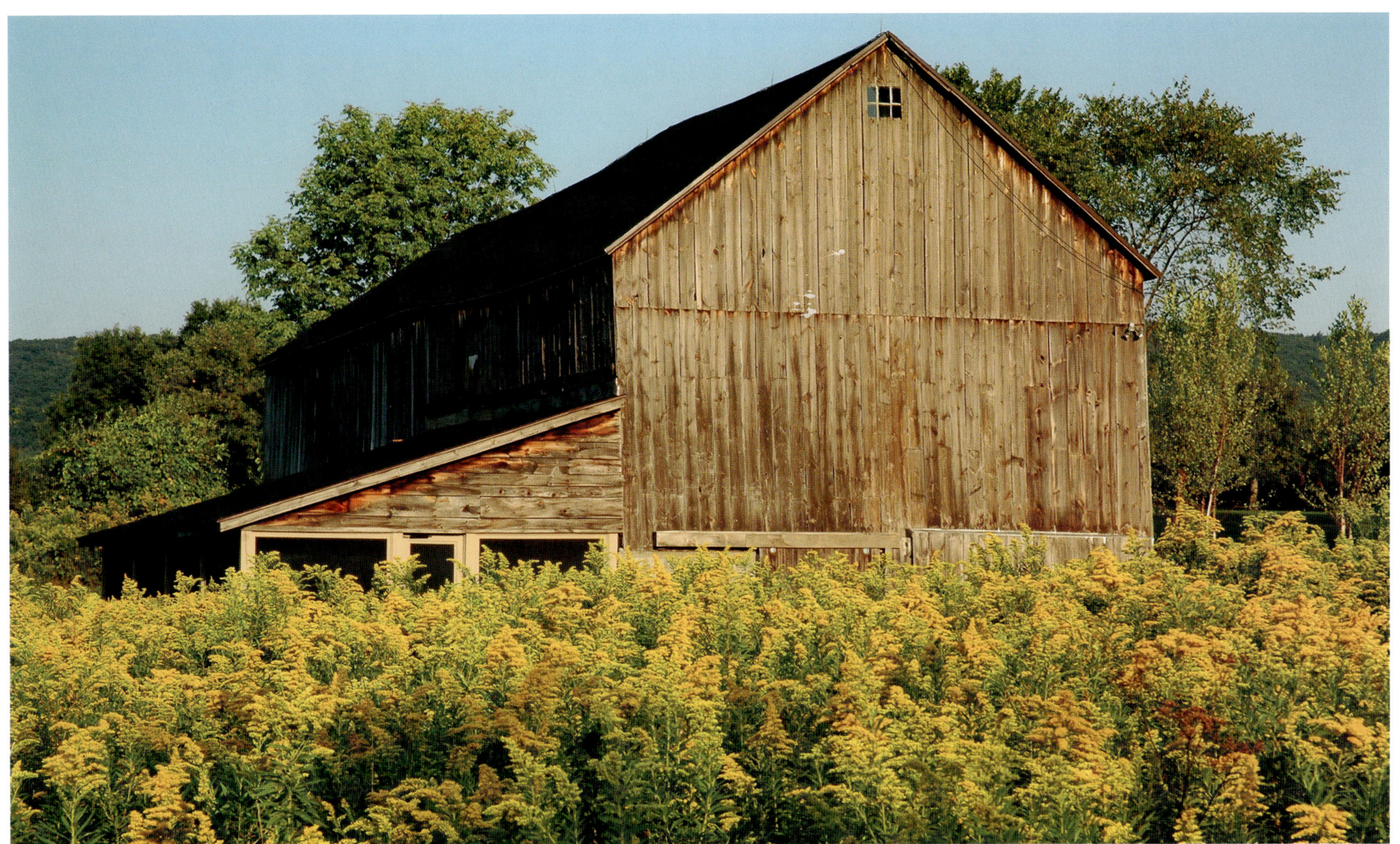

An example of a simple gable vent on a barn in West Stockbridge. Since this was one of my drive-by shootings, no date or construction information is available for this barn.

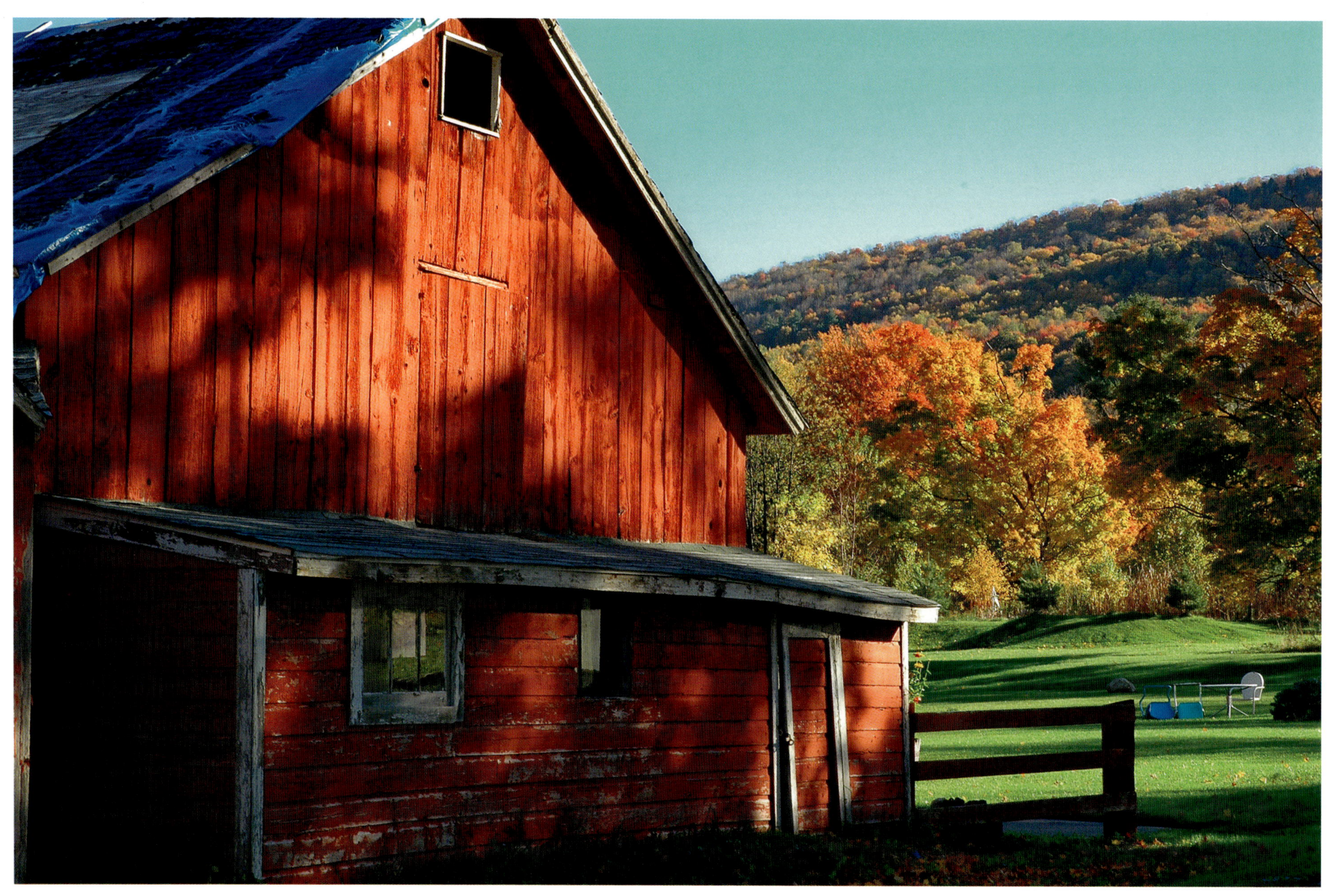

Another example of a gable vent on the smaller 1782 English barn at Golden Acres Farm in Adams. This vent may have been added long after original construction.

Another extremely practical development began to emerge around this time. The traditional hinged double doors of the English barn had proved to be another liability during the winter months. As I have written above, farmers had become sensitive to the need for sunlight and fresh air in the barn for the sake of their livestock. Beyond the addition of a transom window, however, the only way to bring light in was to open up the south facing double doors. In the winter this obviously created the problem of exposing the whole inside of the barn to massive blasts of cold air. In windy weather at any time of year a 6 ft. by 10 ft. hinged door could be unwieldy. An early innovation was the installation of a regular sized door in the face of one of the doors. Eventually, however, it was the introduction of the single sliding door that provided the most practical and enduring solution. Sliding doors began to appear around 1850. They were a common feature on newer New England barns and were very often retrofitted on the older English barns that were still standing and in use.

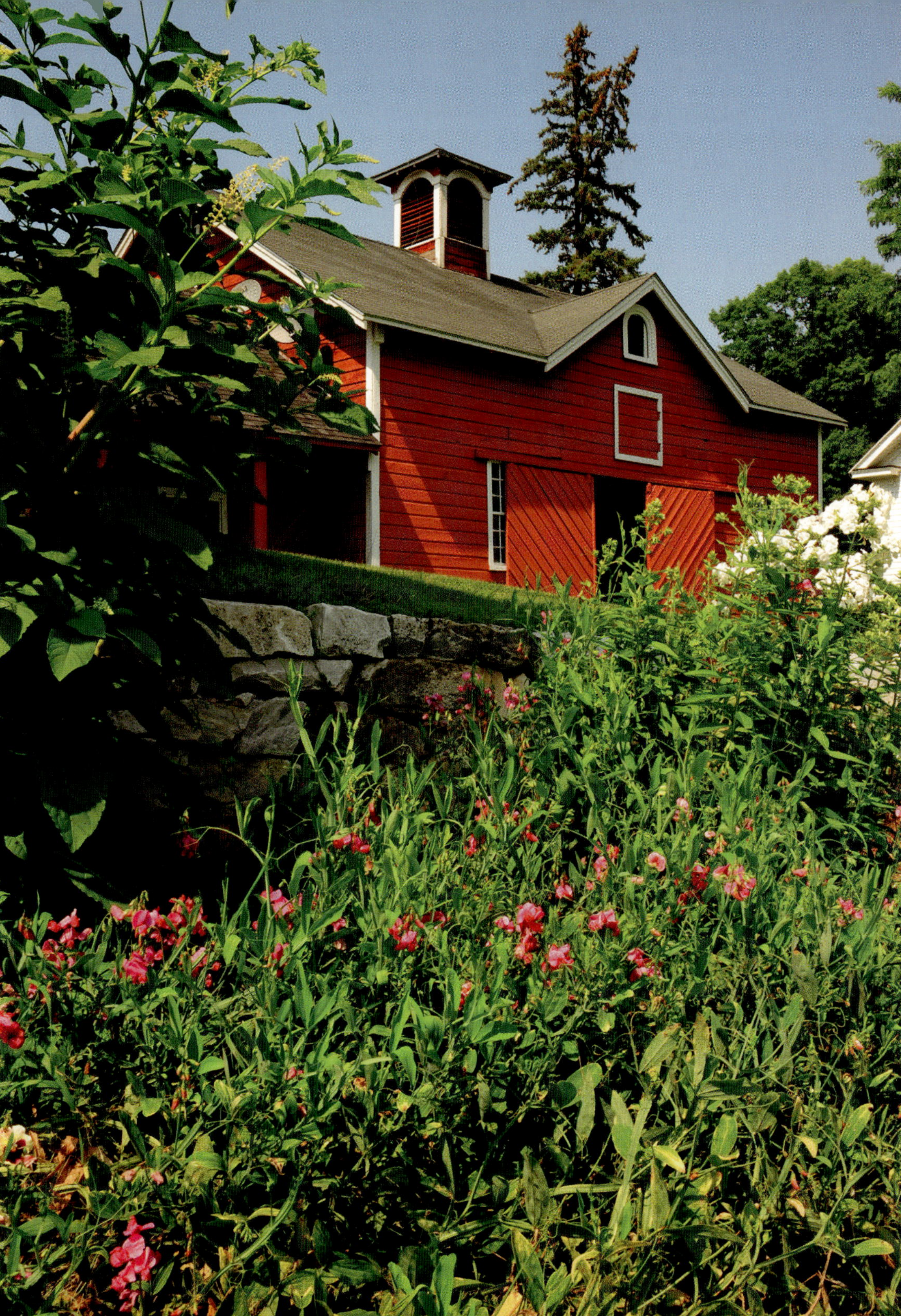

An example of a decorative Italianate cupola on a barn in Sheffield. The barn design, with a gabled wall dormer, is reminiscent of the Gothic Revival style popular in the latter half of the 1800s.

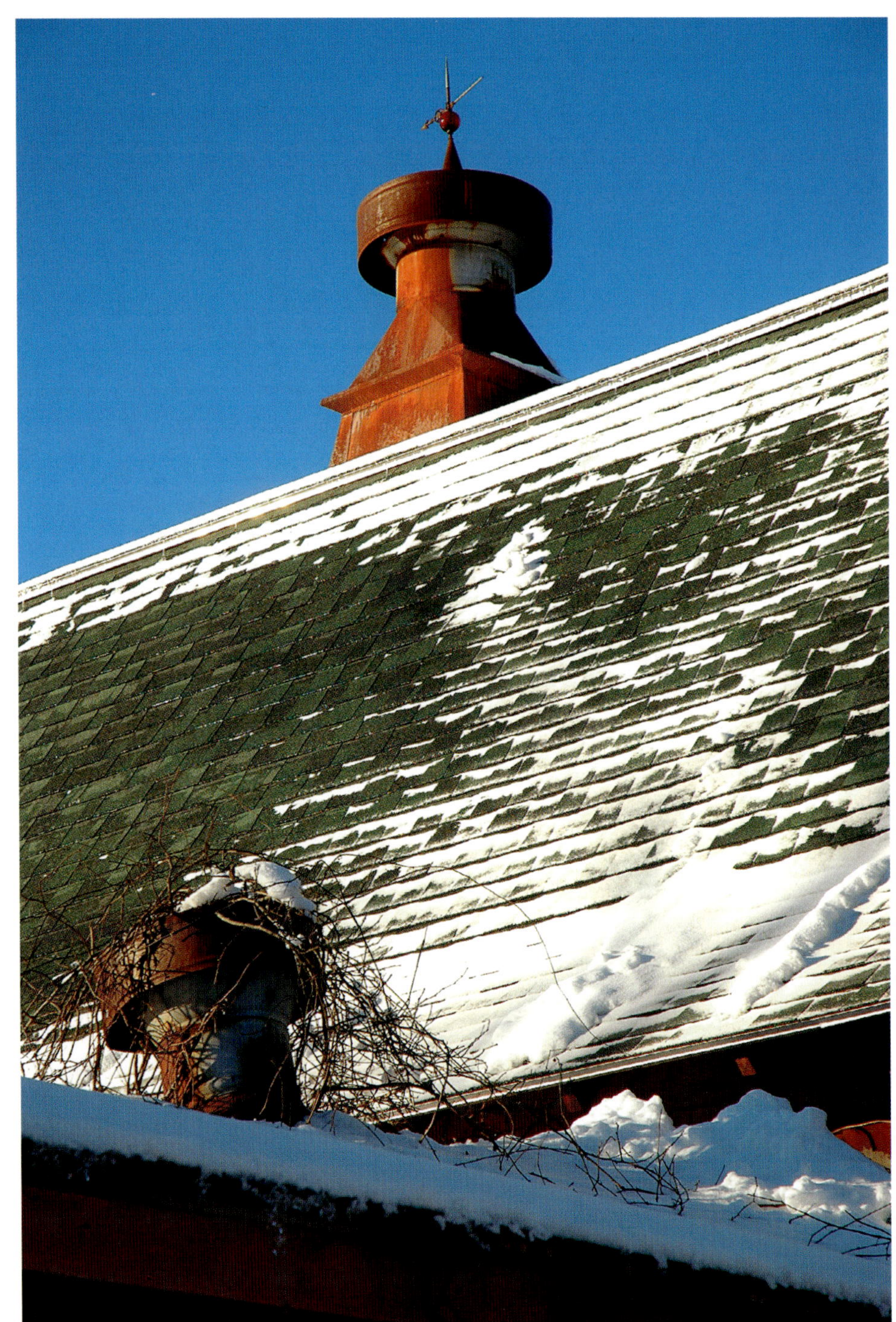

An example of the steel ventilator cupola that was pioneered by the Louden Machinery Company of Fairfield, Indiana.

However, it was not just minor external design elements that were changing in the early 1800s. There was one final fundamental alteration to the English barn that gave birth to its next generation and to a completely American barn form. It was a seemingly simple change—the placement of the main door at the gable end—that created the New England barn.

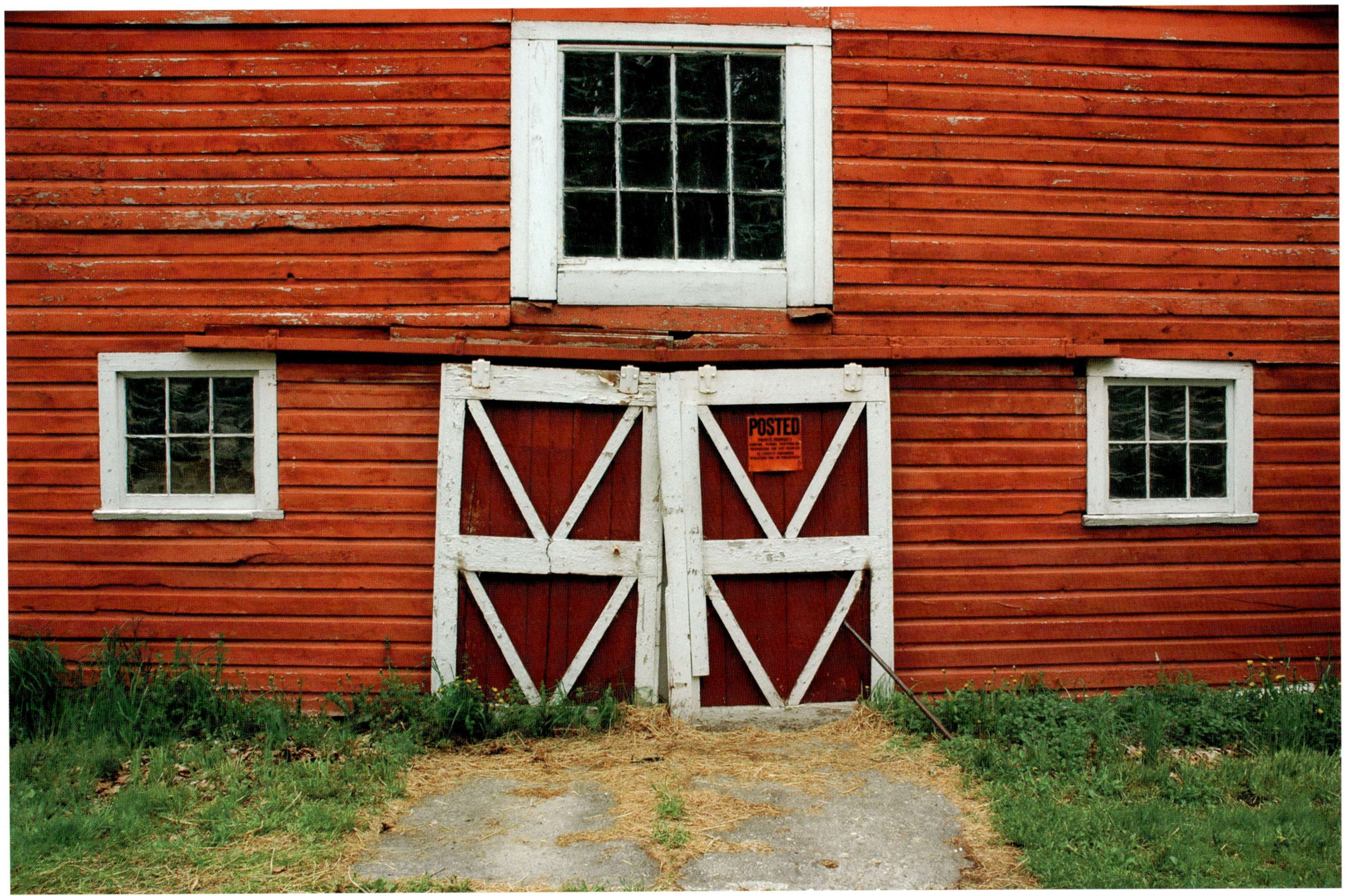

An example of dual sliding doors on this diminutive horse barn in West Stockbridge.

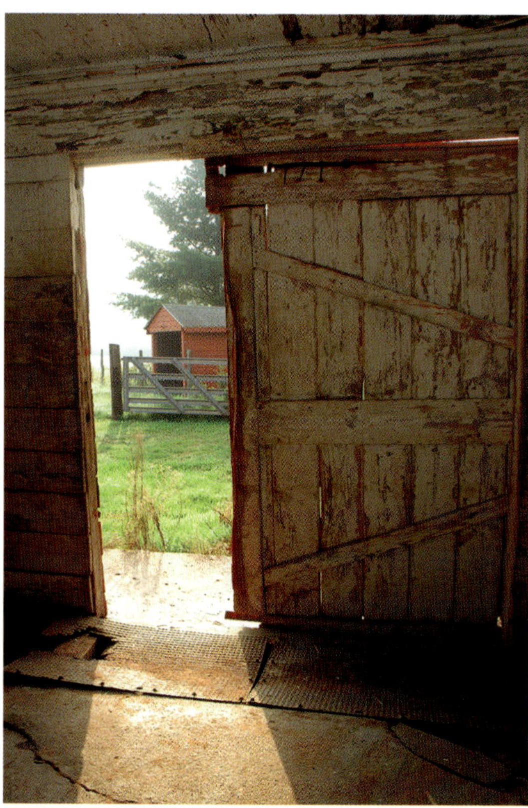

View to the outside through the gable end sliding door on Ron Majdalany's barn in Great Barrington. This entrance is to the milking floor at the south end of the barn. The sliding door was an enormously practical improvement in many ways.

The gable-entry barn was an answer to logistics issues that began to crop up when farmers tried to address the needs of their growing farms by simply adding additional bays to the gable ends of their original English barns. It worked for a while but eventually farmers found it entirely impractical to have to open and close barn doors multiple times and drive in and out of numerous sections of what was essentially the same barn building. Finally the needs and scale of many farms had outgrown the precious, simplistic utility of the original English barn design. By placing the door at the gable end, and often increasing the overall dimensions of the barn, they solved many issues with one master stroke. First, barns regained the advantage of being single drive-through structures. Second, even if the dimensions weren't changed from the 30 x40 model, you could now have two side bays with cattle stanchions running the full 40 ft. length of the barn, increasing the livestock capacity of the same essential structure by a factor of almost three. A barn just a few feet wider and another 10 ft. longer could handle almost 40 head of cattle compared with about 11 in the early adaptation of the original three-bay threshing barn model (similar to Jack Sobon's barn).

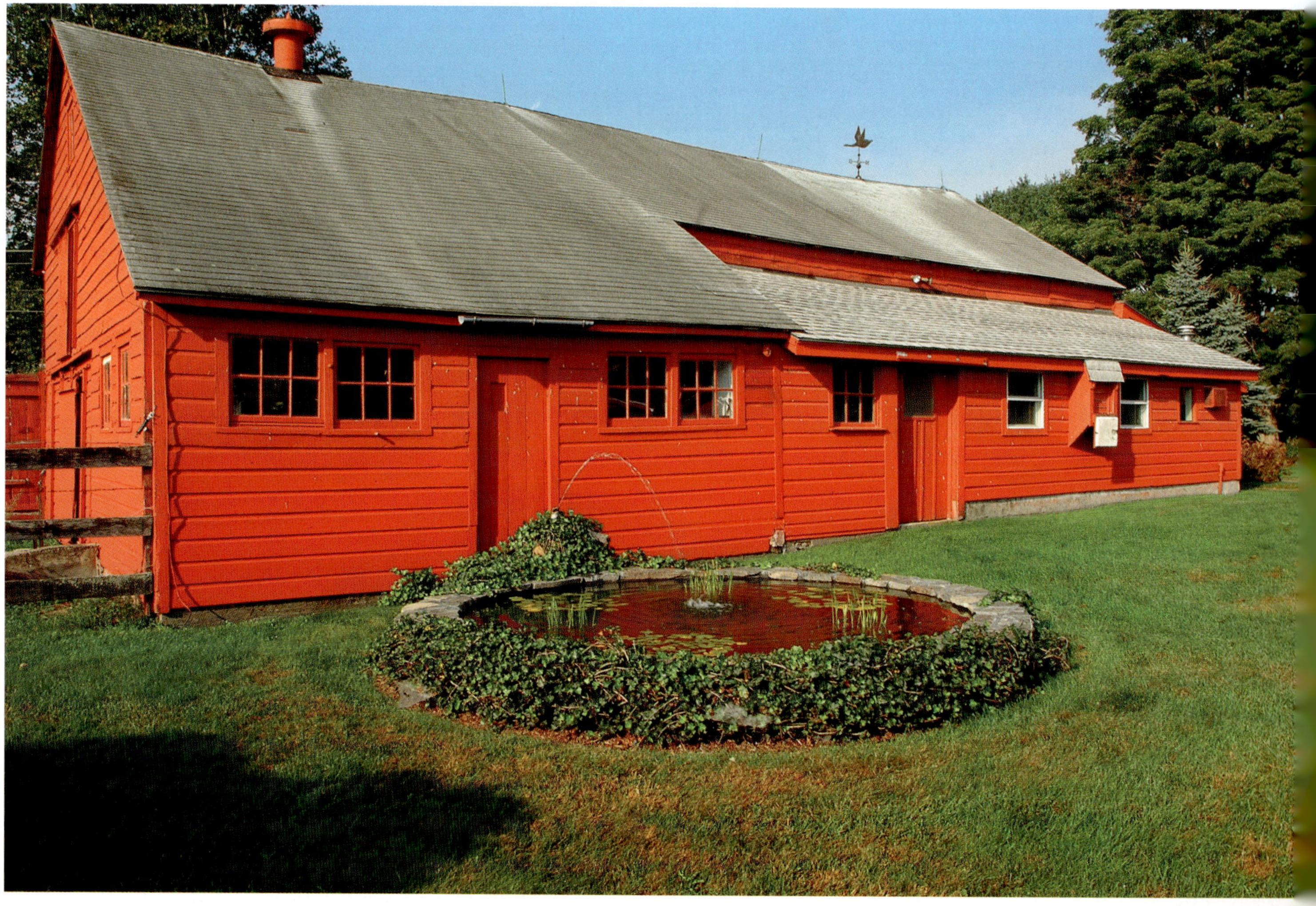

Majdalany's bright red animal hospital is truly one of a dying breed of old English barns but his constant loving care assures it a solid future.

Additionally, whatever the original length of the barn, new sections could now easily be added at either end to increase the capacity of the barn without affecting its drive-through efficiency. The new design was also cheered by farmers because it meant that they no longer had to drive through rain water dripping off the eaves when entering the barn in inclement weather.

Naturally the new design was not without its drawbacks. The eaves of the gable roof now came down over the area above the side stalls that was used for hay and grain storage thus significantly reducing head room and the available storage area. One solution was to make the frame bents of the barn taller, raising the side walls, to regain the height in these mow areas. Another common alternative was to use only one side of the barn for cow stalls, leaving the other side bay, or some significant portion of it, open for storage. In order to maximize exposure to sunlight for animals, the stalls were placed on either the south or east side, depending on the barn's geographic orientation.

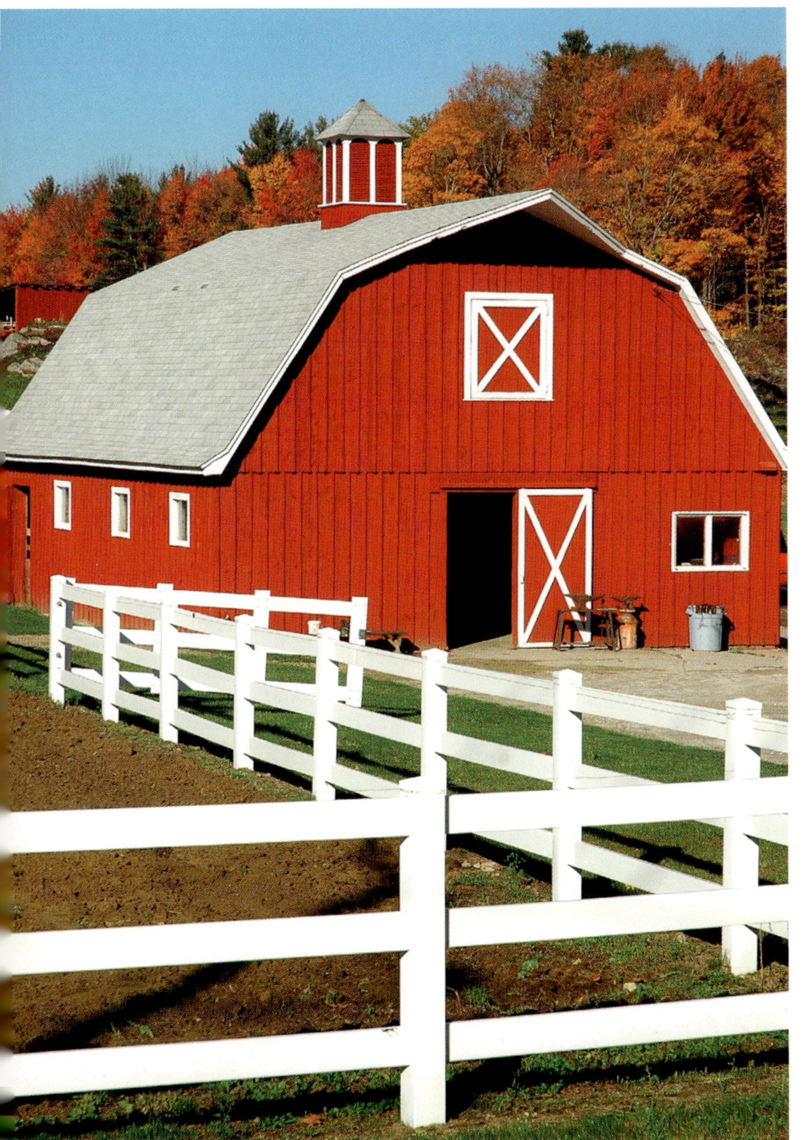

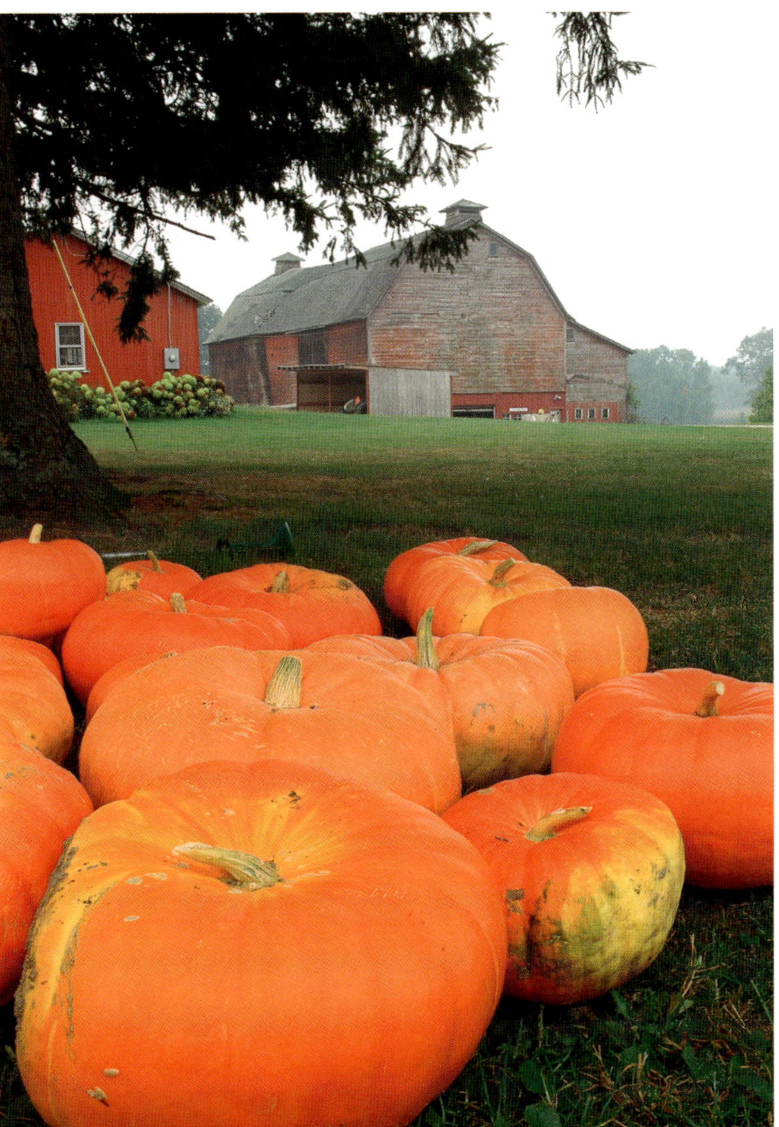

Left:
The gabel-entry gambrel roof barn at Trailside Farm in North Adams is a good example of the size and shape of this style of barn that quickly gained popularity around 1800-1820. The barn exterior features board-and-batten siding.

Right:
The Boardman's gambrel roof barn is dressed up for Halloween.

According to Thomas Hubka, there was one other significant and defining change to the traditional timber frame barn that was introduced around this time. It was the culmination of the many process and technology developments discussed above that came to the forefront in the first half of the nineteenth century. Hubka stated it clearly and comprehensively in his article *The Americanization of The Barn*:

> The process of Americanization and change in barn construction actually began soon after initial settlement and was motivated by the problems and opportunities of farm life in the New World... These small, continuous refinements in barn making erupted in a great change in barn building during the late nineteenth century and created a watershed division in American barn construction between modern and premodern barn types. The new barn that crystallized this major evolution was the Gambrel (double sloped roof) barn. It was not only the shape of the roof that made it new, but vast changes in the building system separated it from previous barns. The Gambrel roof barn incorporated standardized, lightweight, machine-sawn structural members into an advanced truss configuration with nail construction. Its new roof truss system produced the modern symbol of the technologized barn, the ubiquitous double sloped Gambrel-roofed barn. Today, more that a century after its introduction, it is the most widely accepted symbol for the American barn. For late-twentieth-century Americans, it is difficult to imagine this now familiar symbol of the farm as a once new and revolutionary structure that transformed barn construction—but that is what it did. The Gambrel barn is the fitting symbol for the American barn, not only because it is so widespread, but because it marks the triumph of the modern technologically advanced Americanized barn.
>
> The Gambrel barn and related barns of the late nineteenth century mark the advent of the standardized construction systems, mass produced building materials, mail-order planning and distribution, and national barn-building traditions. These vast changes did not happen swiftly, but were incrementally applied by individual farmers over long periods of time. Within all periods, new ideas competed with older existing traditions of barn building. For example, even the adoption of a new gambrel roof system with stud walls and a truss roof did not wholly eliminate the old heavy timber mortise-and-tenon construction system. Barn builders frequently integrated both old and new systems into the overall structural framework.[13]"

In New England and the Berkshires, in particular, the gambrel roof design helped farmers solve the problem of lost storage capacity posed by the introduction of the gable-entry New England Barn. By increasing the pitch of the lower portion of the roof considerable space and head-room were recovered in the side bay areas. Inexorably, issues of technology, practicality and the scale of farming operations were making the original English three-bay threshing barn obsolete.

Right:
This barn on Cooper Hill in Ashley Falls is similar in shape, size, and scale to barns that can be found throughout the Midwest.

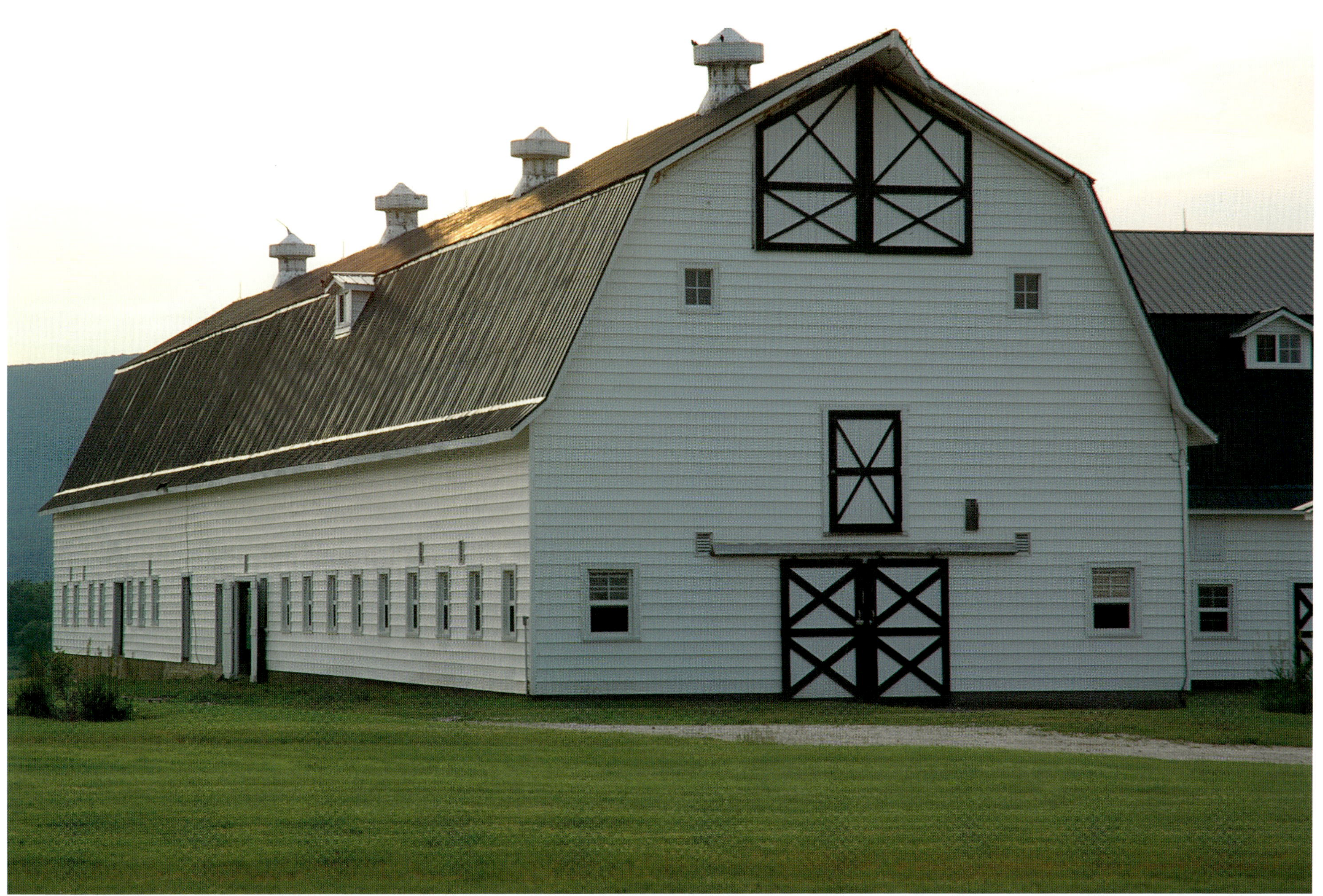

Despite the fact that it was gradually being phased out, it is comforting to know that the English Barn, the great ancestor of so many barns of Berkshire County, did not become completely extinct. It did, however, evolve. The core genetic makeup of the English Barn, the essential post and beam frame design, was carried west with the pioneer movement by both the descendants of English settler-farmers and the newer British arrivals. According to *The Old Barn Book*, timber frame post-and-beam descendents of the English Barn have been found in such places as Ohio, Indiana, Wisconsin, and even Utah, and some are still standing today[14]. Many of them are much larger versions of the original 30 x 40 three-bay design, but retain all of its core characteristics. And, of course, many English barns that may have simply deteriorated, collapsed and completely vanished have been given new life and purpose as interest in barns saw a resurgence over the last 20-30 years. This has defined the work and careers of people like Jack Sobon and Dave Lanoue who continue to apply craftsmanship and a deep knowledge of technique to barn restoration and relocation projects throughout the United States.

In the shadows of this explosion of timber frame barn building techniques, materials, designs and technology in the early-to-mid 1800s Berkshire County became the birthplace of an entirely new and unique barn form. In 1825 a fire destroyed a large dairy barn on the Shaker community in Hancock. The community had grown rapidly in the early 1800s and, having developed a diverse portfolio of industries, was quite prosperous by the time of the fire. The dairy business was a significant part of this enterprise and the community elders knew that they needed a replacement barn as soon as possible. What resulted, however, was the enigmatic round stone barn. The barn is a highly complex structure and it defies every convention of Shaker architecture that had been established by that time. Shaker architecture had been characterized by clean elegant lines, austere appearances, straight geometric shapes, and exacting symmetries.

Yet while it radically broke the mold, it still managed to embrace many of those essential elements—a circle is infinitely symmetrical on its diameter, and when viewed from ground level the barn has a very well tailored wedding cake look. But the most intriguing aspects of the project were that the community fathers did not hire an architect to design the project, no plans or blueprints have ever surfaced, and they managed the project themselves. Furthermore, this type of building was almost entirely without precedent. The only other round barn known to exist at the time in America, which was actually a 16-sided polygon, was a timber frame building George Washington had commissioned on his Virginia estate. His barn was completed in 1794, and although there are many similarities in design, there is no written record of any kind that suggests that the Shakers took this building as their initial inspiration. Since there are no written records of any phase of the project, however, we are left with mostly theories, conjecture, and best guesses regarding planning and construction—just as had been the case for historians and researchers trying to solve the mysteries of the early English barns.

Eric Sloane called the round barn building tradition that the Hancock barn spawned the first American "modern architecture."[15] He believed that the Shakers saw the circle as a utopian symbol. Since the circle was employed in many aspects of their lifestyle and group organization—their activities included sewing circles, praying circles, and singing circles, and many of the objects and artifacts they produced were round—he believed that they may have built a round barn as a permanent monument to the purity of their lifestyle[16]. It may, of course, have had nothing to do with creed or social philosophy. It may simply have been that William Demming and Daniel Goodrich, the community leaders who oversaw the project, fused whatever they knew to be the best technology, construction methodology and materials, and architectural innovations with the need for maximum efficiency in their operations and came up with their own unique concept. What we do know for certain, just as was the case with the barns of the English settler farmers 100 years before it, is that the Hancock Shaker round barn would never have been built without the ingenuity, intellect and determination of its creators. The barn required a small army to build and its $10,000 cost was considered extravagant by many in Berkshire County at the time, but it has come to be one of the most revered and visited attractions in the County.

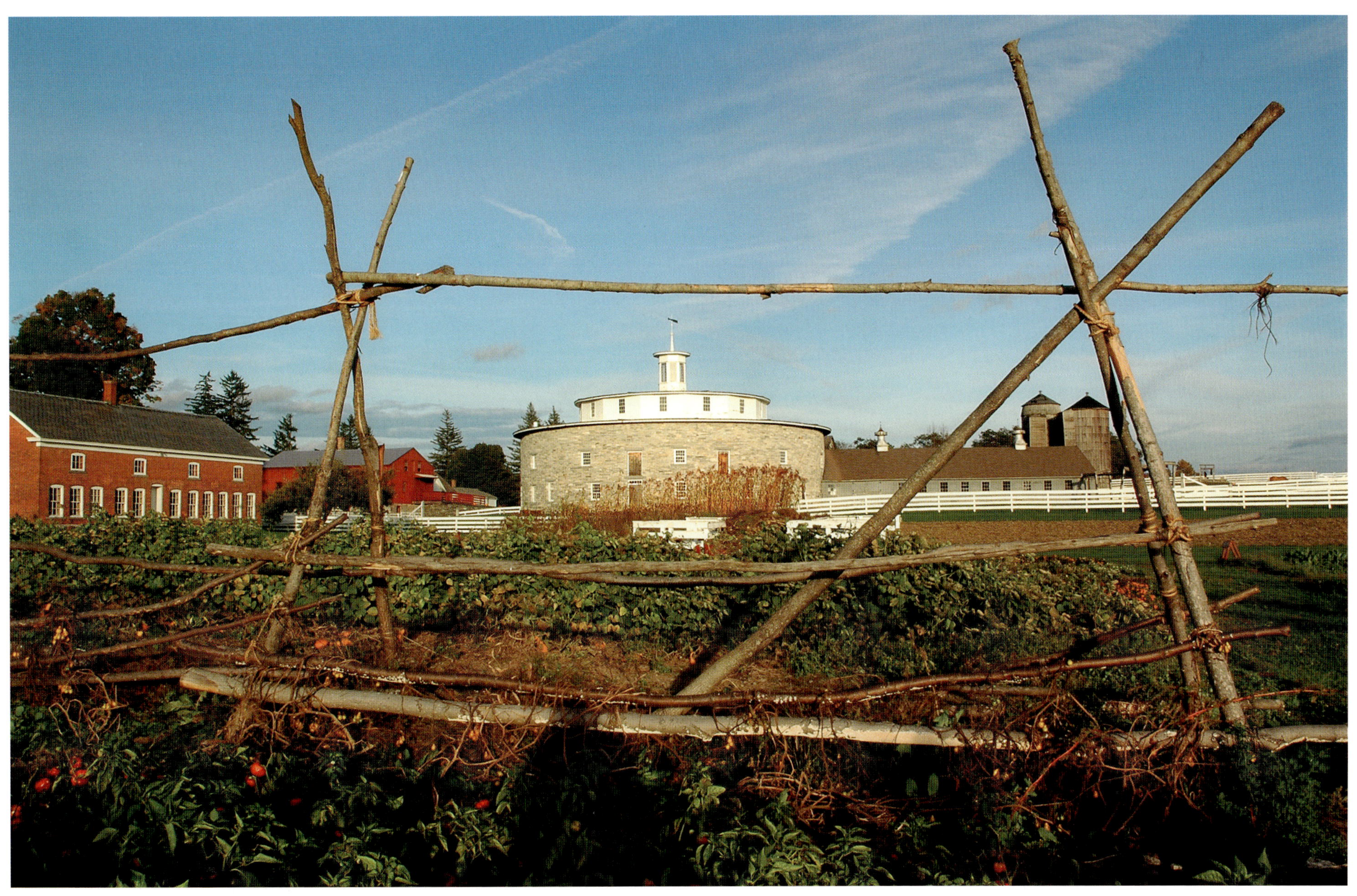

The Round Barn at the Hancock Shaker Village was a truly innovative masterpiece of design. It was revolutionary in its time and sparked a widespread building spree of round barns that spread in every direction despite the fact that it was, at first, considered an ostentatious extravagance by many.

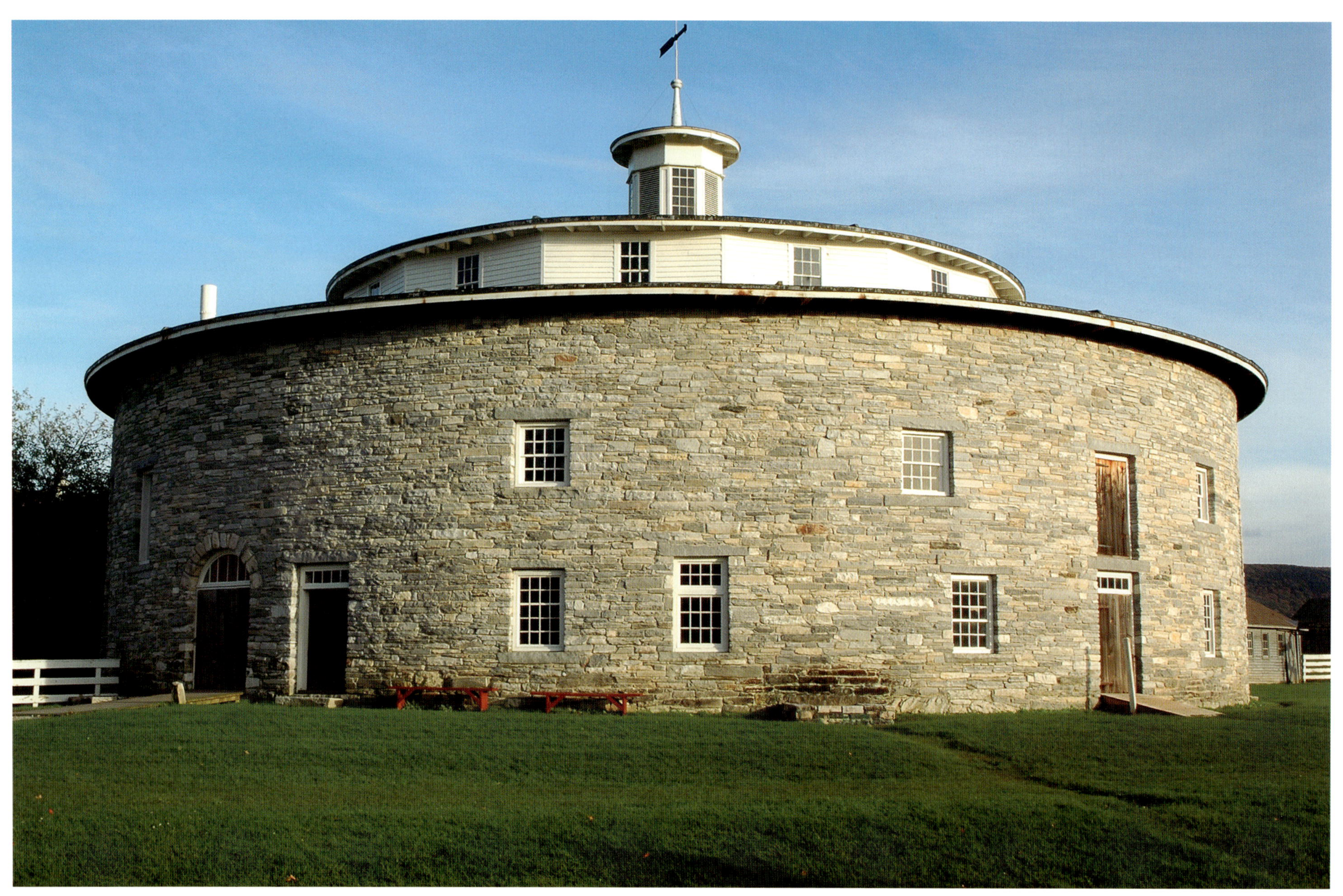

With its elegant lines, strong demeanor, and wedding cake tiers, the Shaker Round Barn continues to attract tens of thousands of visitors every year.

There were countless other modifications to barns—too many to cite here—so I will only list a few that can be recognized on barn exteriors in the Berkshires.

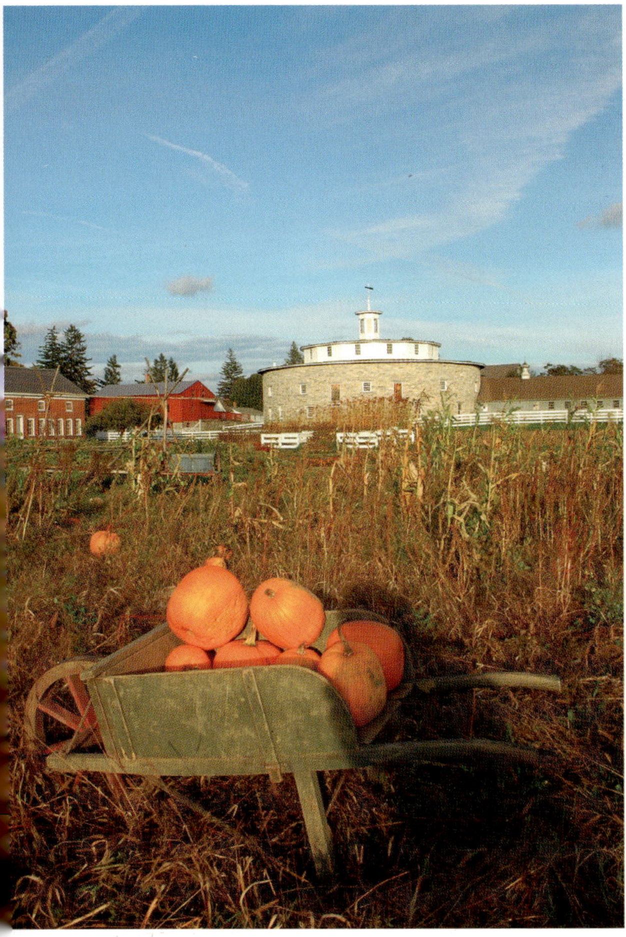

Harvest time at Hancock Shaker Village.

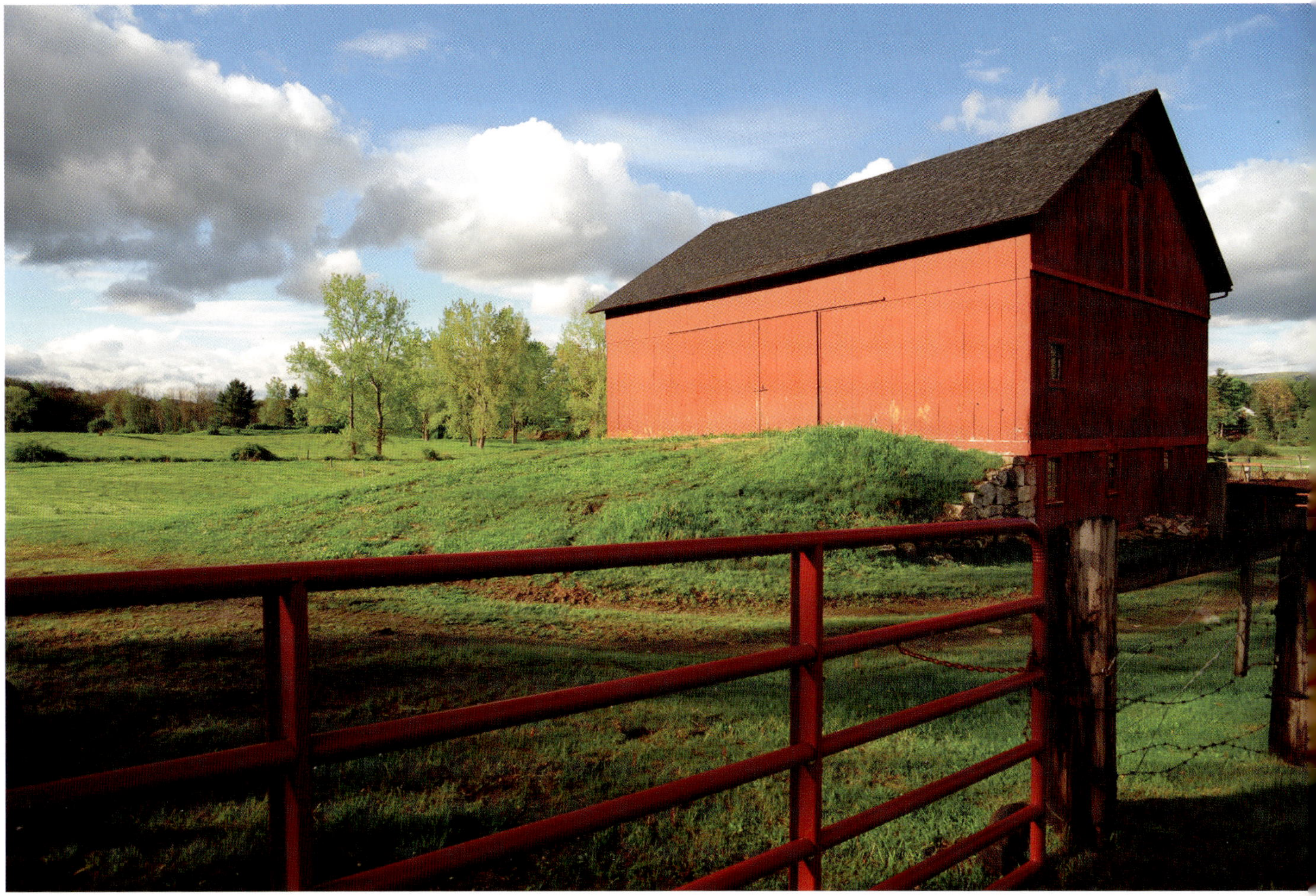

This English bank barn on Division Street in Great Barrington is an 1875 rebuild of the original 1820 barn that was destroyed by fire. The replacement barn was built on the original foundation.

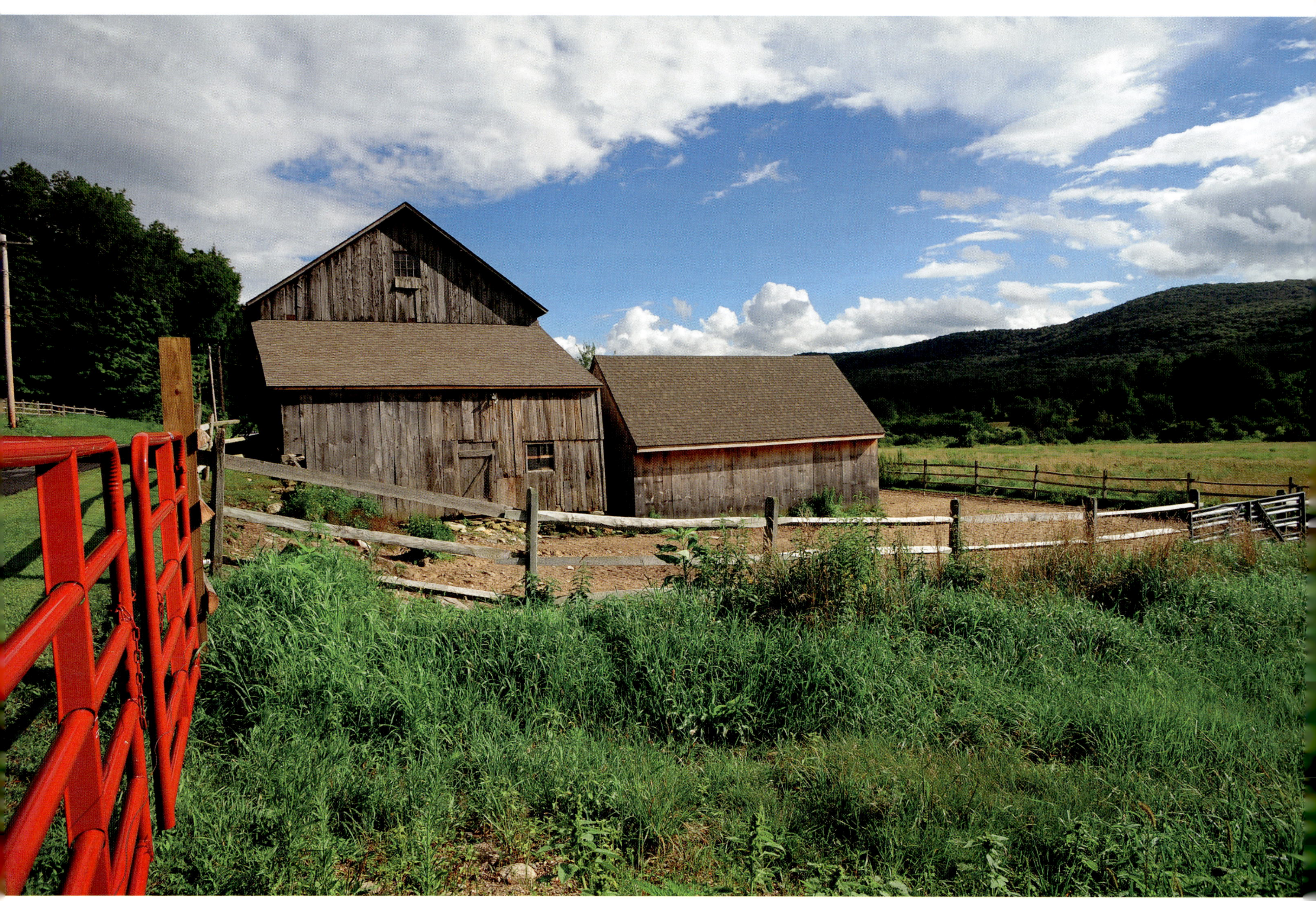

A side-entry English bank barn in Tyringham. Date of construction unknown.

In addition to the gambrel roof, new concepts affecting the location of barns were adopted to address the quandary of lost storage space that came into play with the rise in popularity of the gable entry New England barn. Various permutations of bank barns, for example, which offered even greater access to all areas of the barn, began appearing by the 1850s, but most types are rare in Berkshire County today. Beginning with the side-hill English barn, new designs were continuously being introduced including reconfigured English barns, gable-front bank barns, high-drive bank barns, covered high-drive bank barns and late nineteenth-century eaves-front bank barns. In these barns there were two, or more, separate levels allowing the cow stalls to be located below the main barn area at ground level. Doors were placed either on the gable or eaves side of the highest level of these barns so that carts could still access the upper storage area. Chute systems were built into the drive-in level floor so that hay could easily be dropped down to the lower level stalls. All of these types of barn were perfectly suited to the undulating topography of the Berkshires.

In Great Barrington, David Leavitt, an exceedingly wealthy New York merchant, built the colossally opulent "Cascade Barn" just north of town at what was then an enormous sum of $20,000. It was part high-drive bank barn, part fortress, and part palace with a waterfall emerging from the middle of its lower stone-wall foundation level for good measure. If any barn in Berkshire County had been built to withstand the ravages of time, weather and human folly, this was it. Ironically, nothing remains of it today except a few drawings and photographs. But that will remain a story for another book to tell.

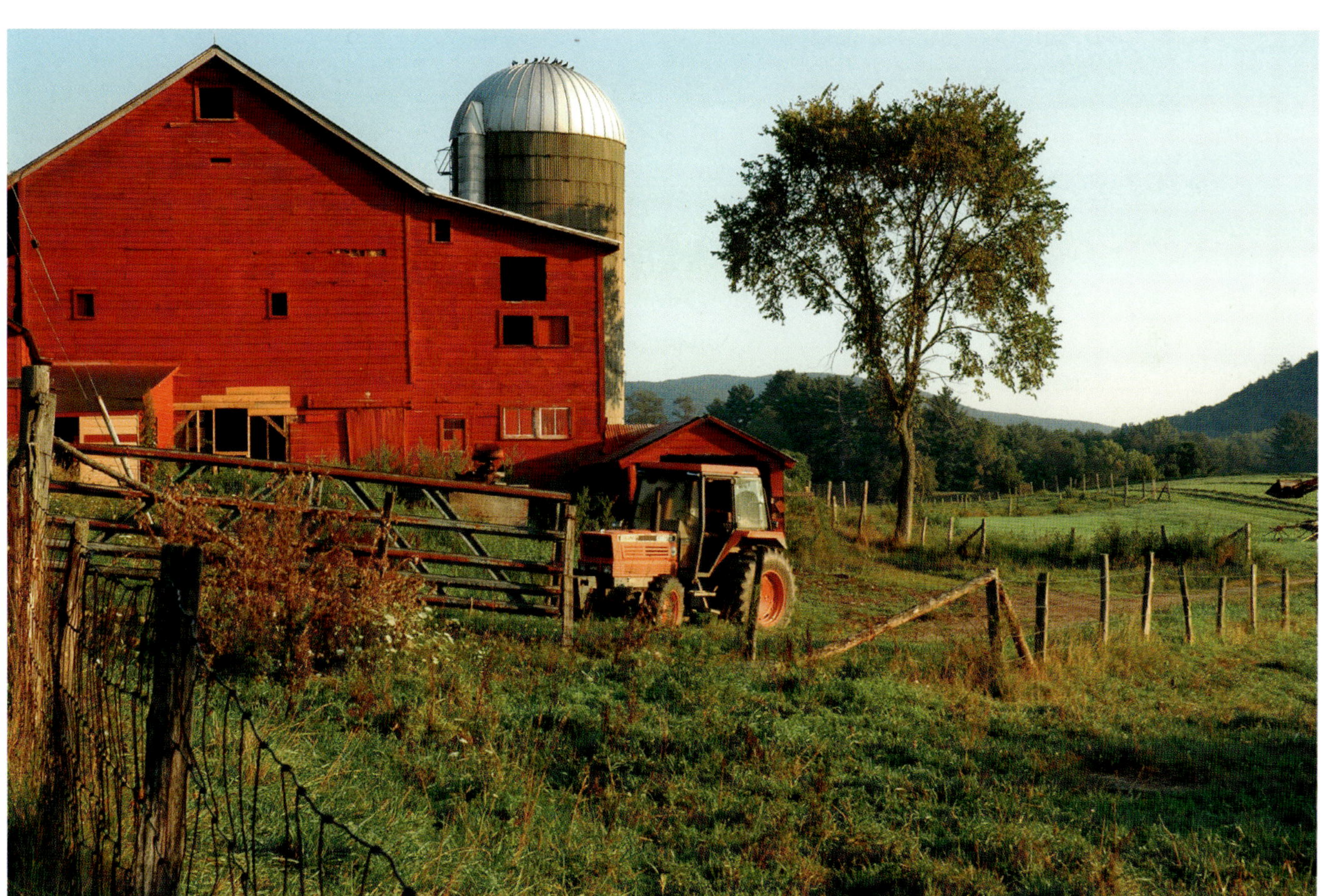

A side entry English bank barn with an addition on the eastern end, c. 1830.

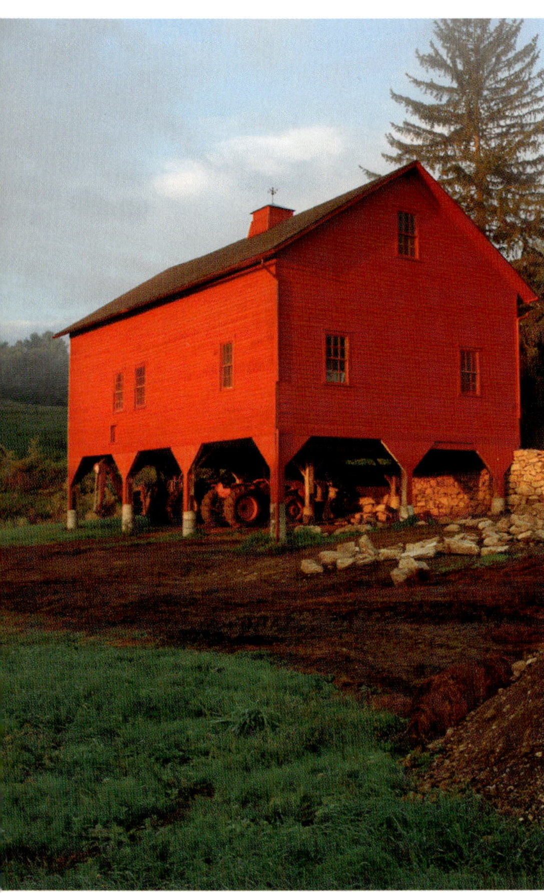

The raised bank barn at Lila's Sheep Farm has been methodically restored and shorn up to ensure that it will last another century, at least. This is the barn for which the foundation wall was rebuilt in 2007.

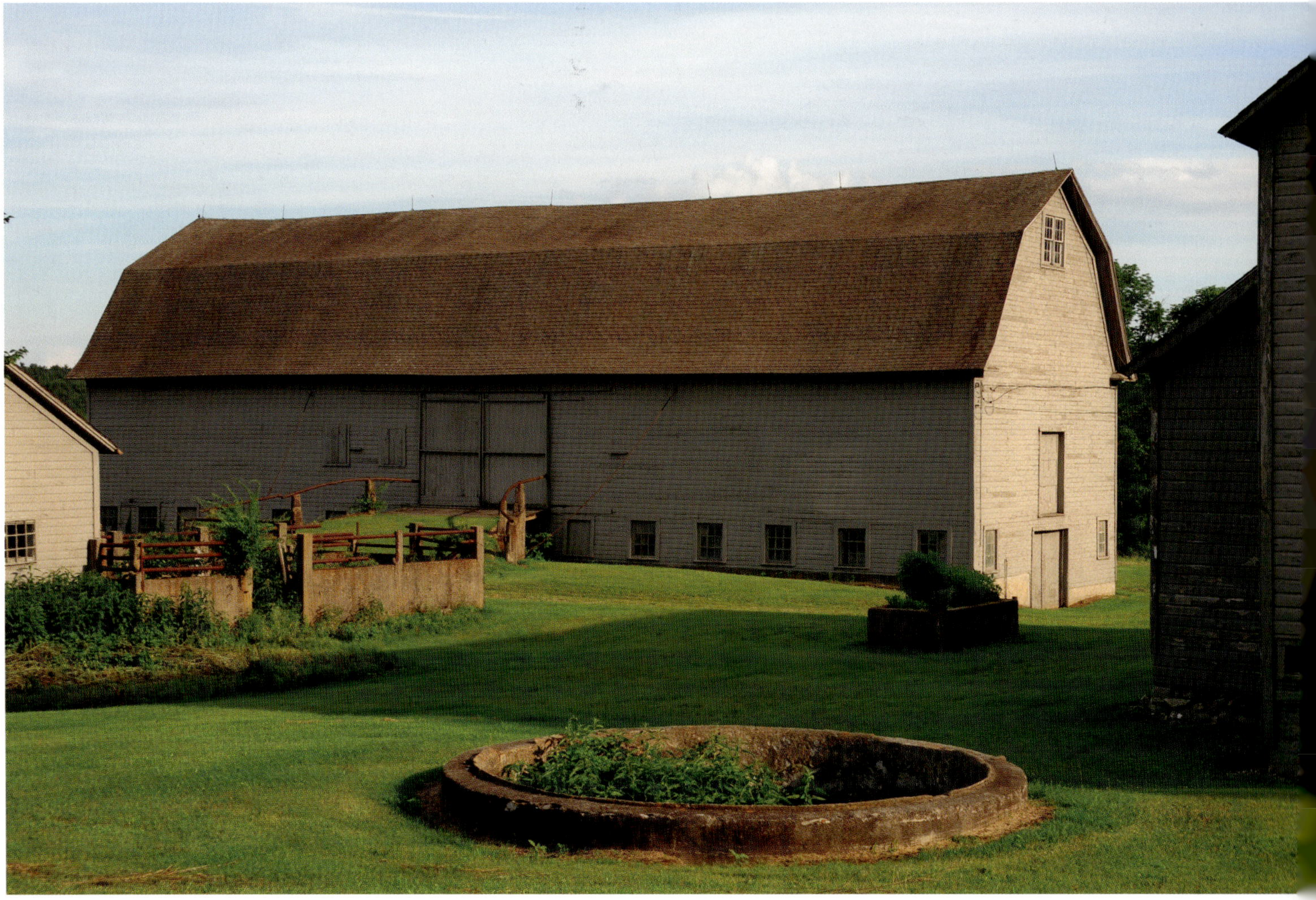

A large gambrel roof bank barn on Route 41 in Sheffield. In the foreground you can see the foundation for what was probably a stacked concrete silo.

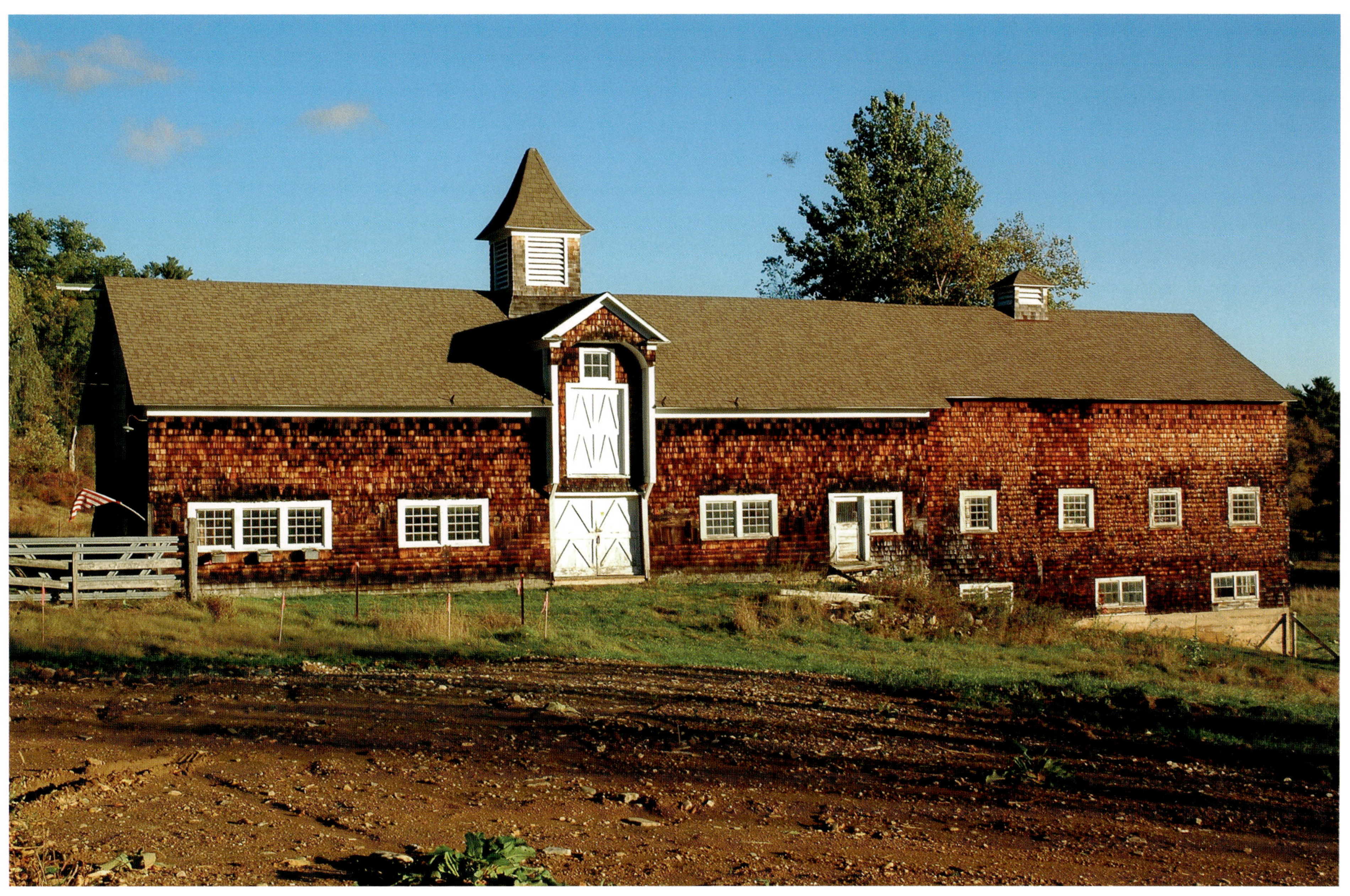

An elaborate shingled bank barn with Gothic Revival accents that probably dates from 1890-1920. This barn is located on Route 57 in New Marlborough.

Beginning in the late 1800s the use of hay tracks installed under the roof ridge led to the placement of haymow doors in the peak of one or both gable ends. The New England Connected Barn (also known as continuous architecture), where several farm buildings, including the house, were linked together in a chain, became commonplace in Northern New England, and occasionally in the Berkshires after 1800. However, they were discouraged, even prohibited in many places, because of fire risk.

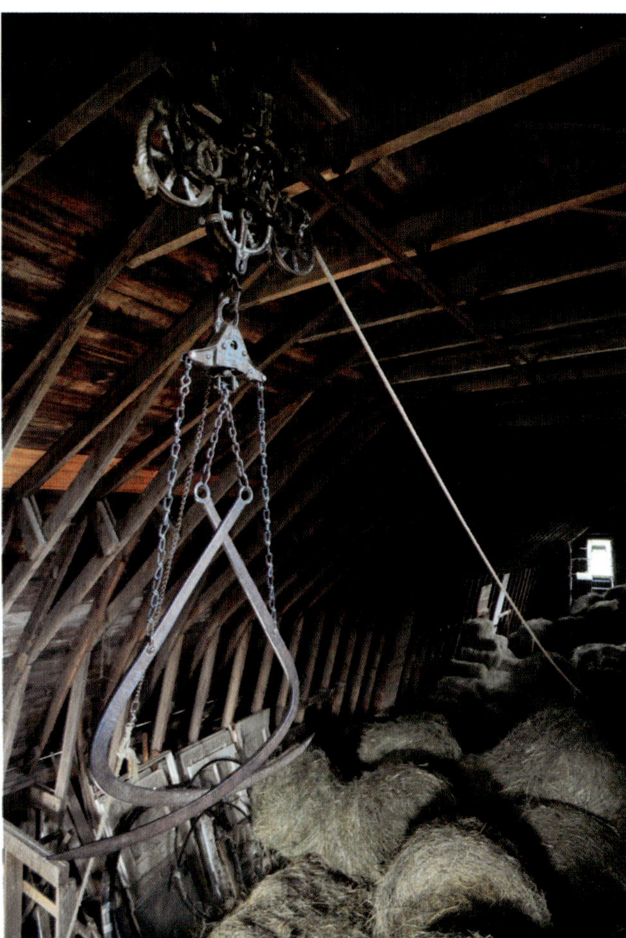

The original hay fork installed in the Leahey Farm barn when it was built in 1934.

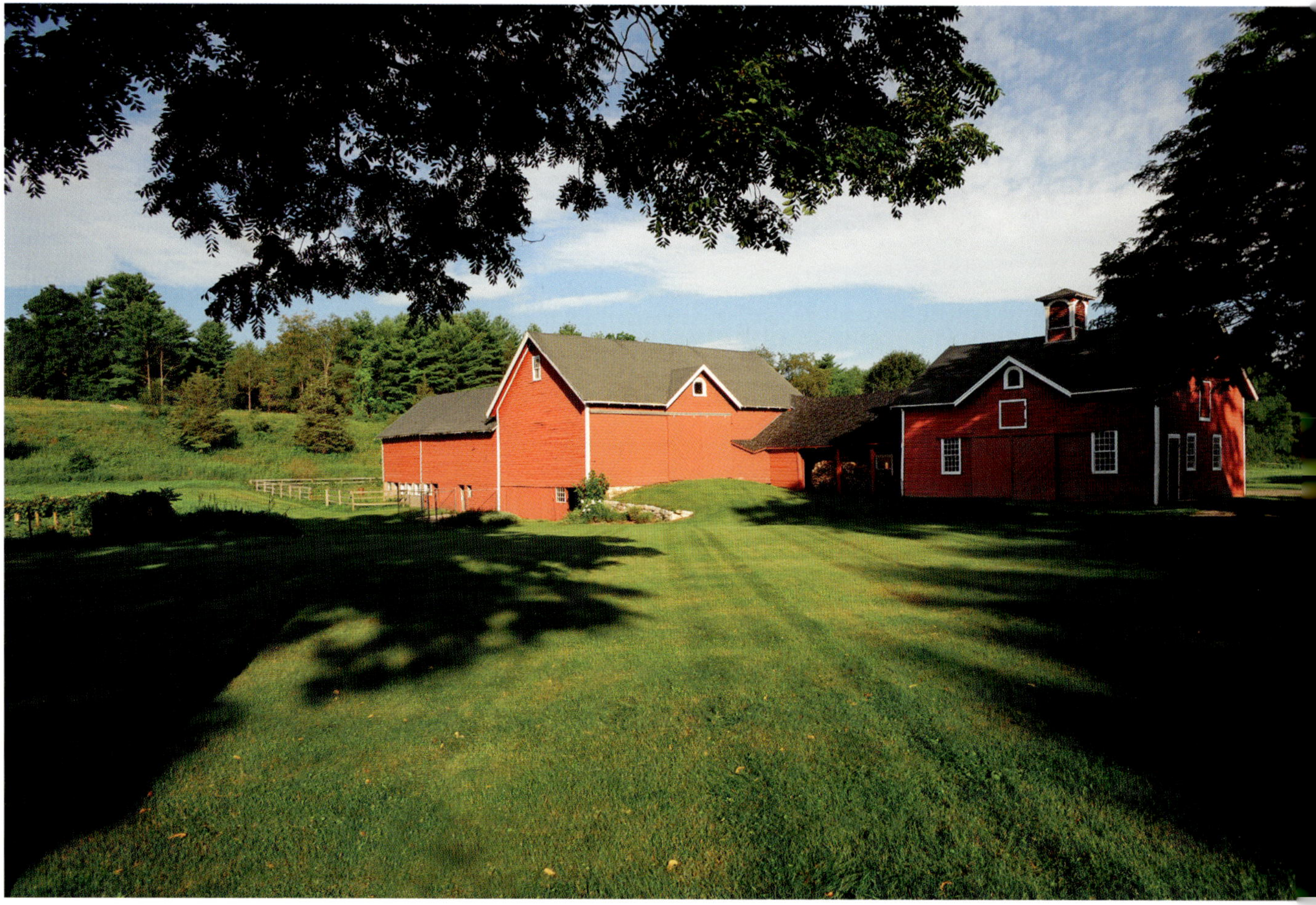

An example of continuous, or connected, architecture, although not in the classic big house, little house, back house, barn configuration. The front barn probably served as a carriage house, or a haymow, while the back two were used for dairy operations.

Right:
The 1892 barn at The Wheeler homestead in Great Barrington and the silo that was added in 1907 and incorporated into the side of the barn. The top of the silo was completely blown off in 1995 by winds from a tornado that went through just a few hundred yards north of the compound.

Silos, another iconic symbol that we tend to associate with every barn, actually did not begin to appear on the exterior of barns until the late nineteenth century. They evolved from the earliest known type of barn, described in the Pre-History chapter. In their earliest incarnations they were known as granaries and corn cribs. Later farmers incorporated cellars under their barns for the purpose of grain and crop storage, and eventually began designing interior silos. Early external silos were square and made of wood. Rare stacked wood silos, and more the common polygonal wooden silos, followed that, and by 1900-1905 wooden stave silos were being built. Precast concrete silos made of stacked cylindrical sections with various styles of roofs were next and by 1950 the concrete stave silo began to appear[17]. Examples of many of these are noted in the photographs.

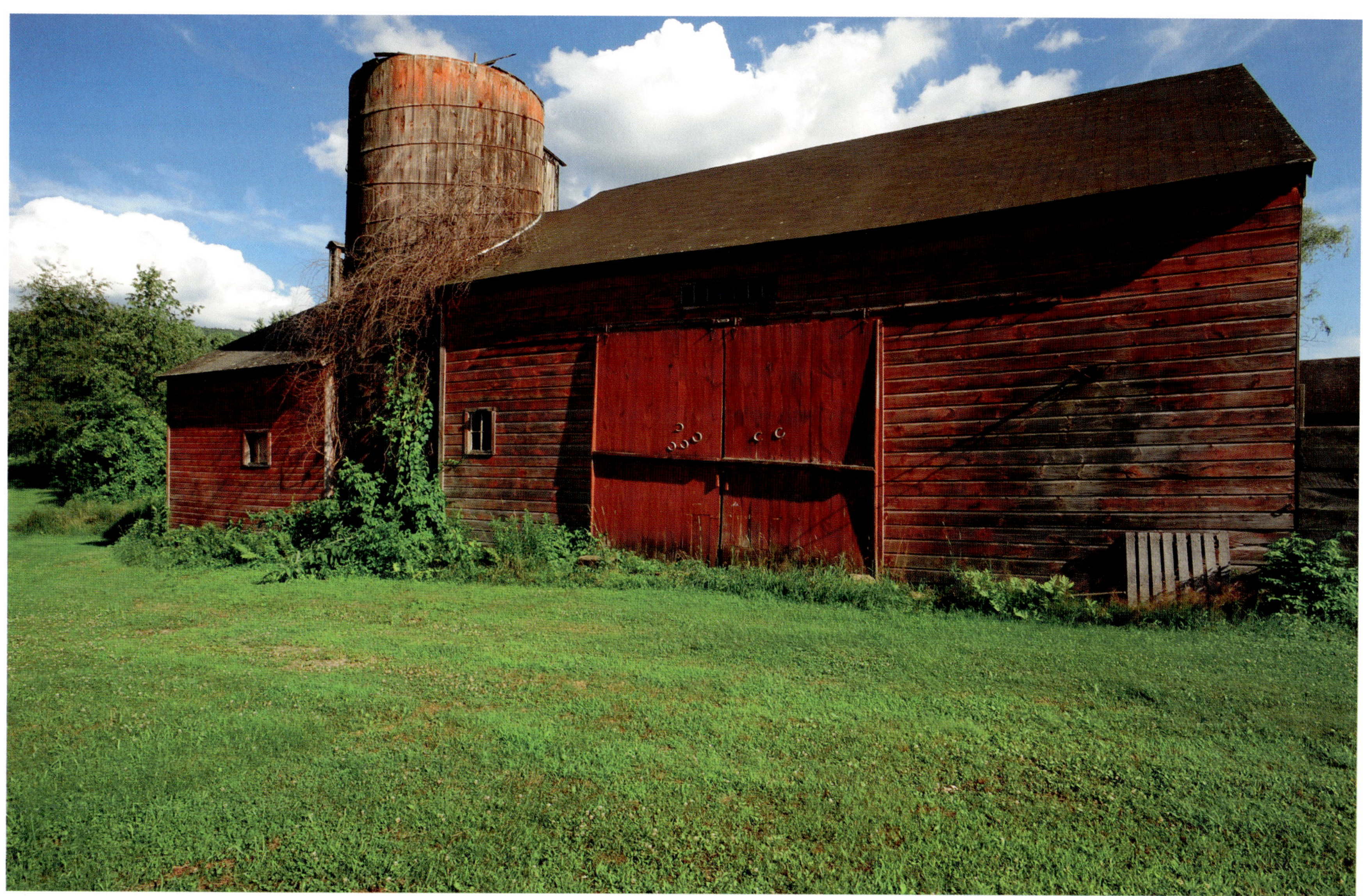

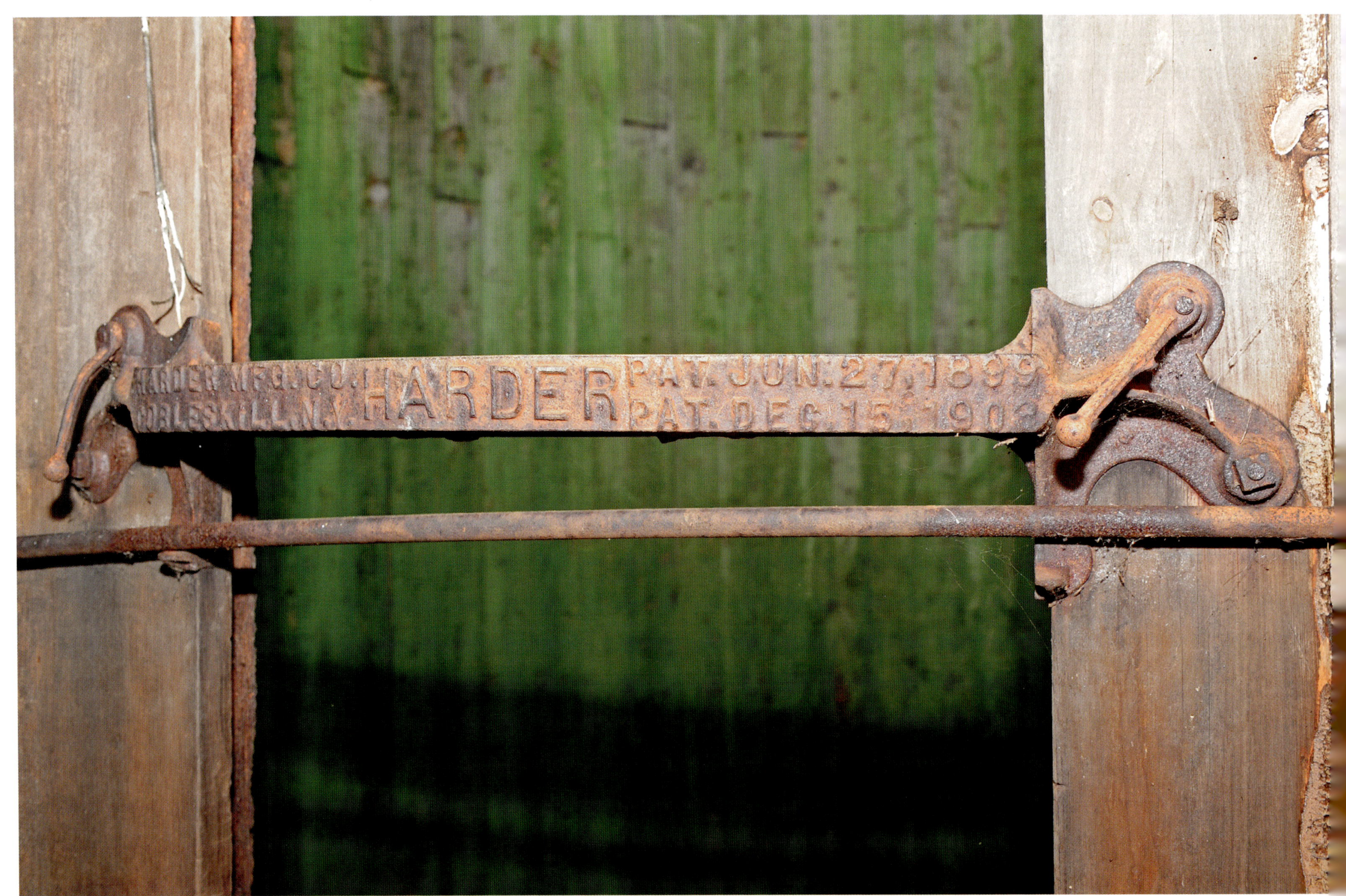

Hardware on the inside of the wooden stave silo at the Truman Wheeler House.

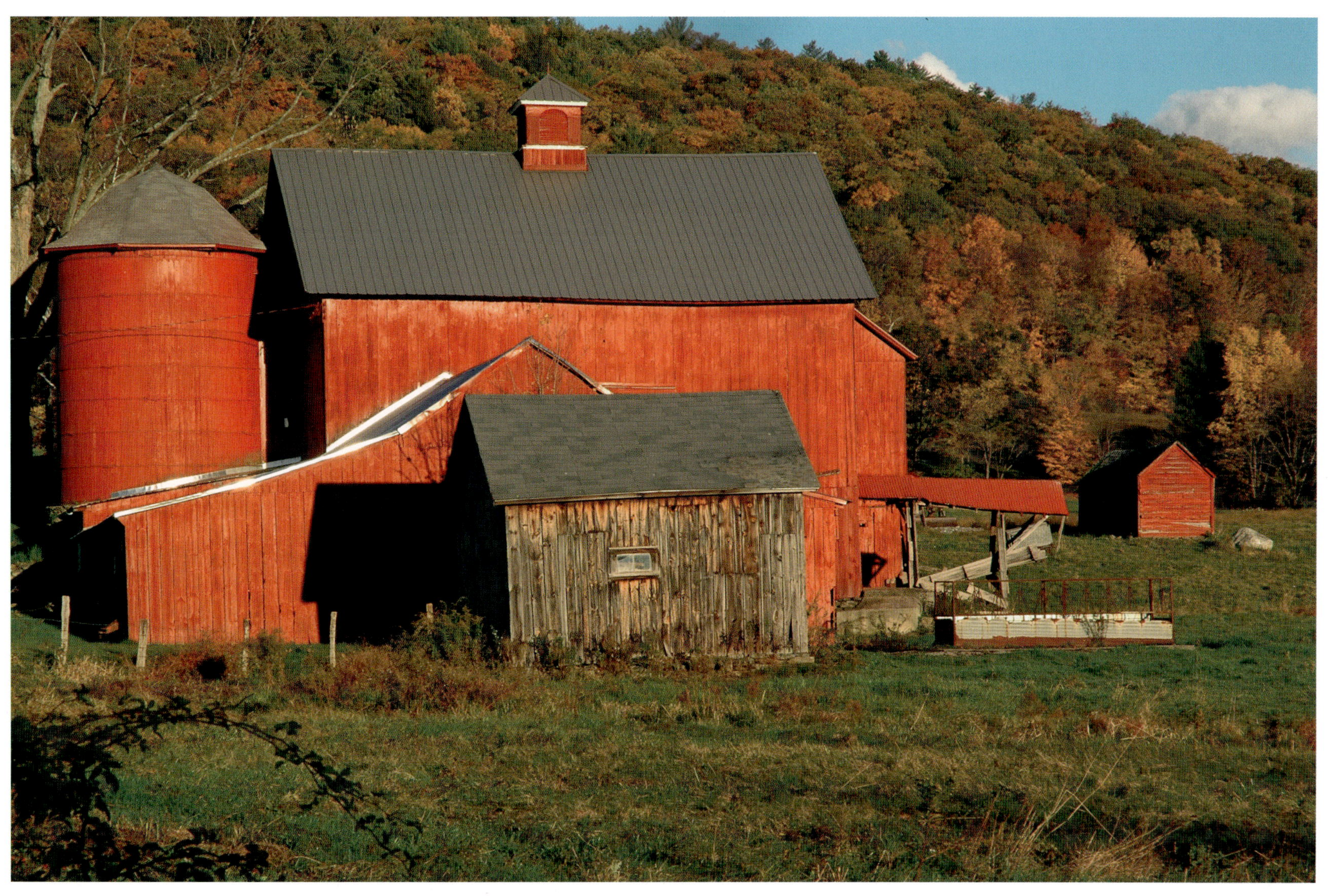

A barn on Green River Road in Alford with a detached wooden stave silo.

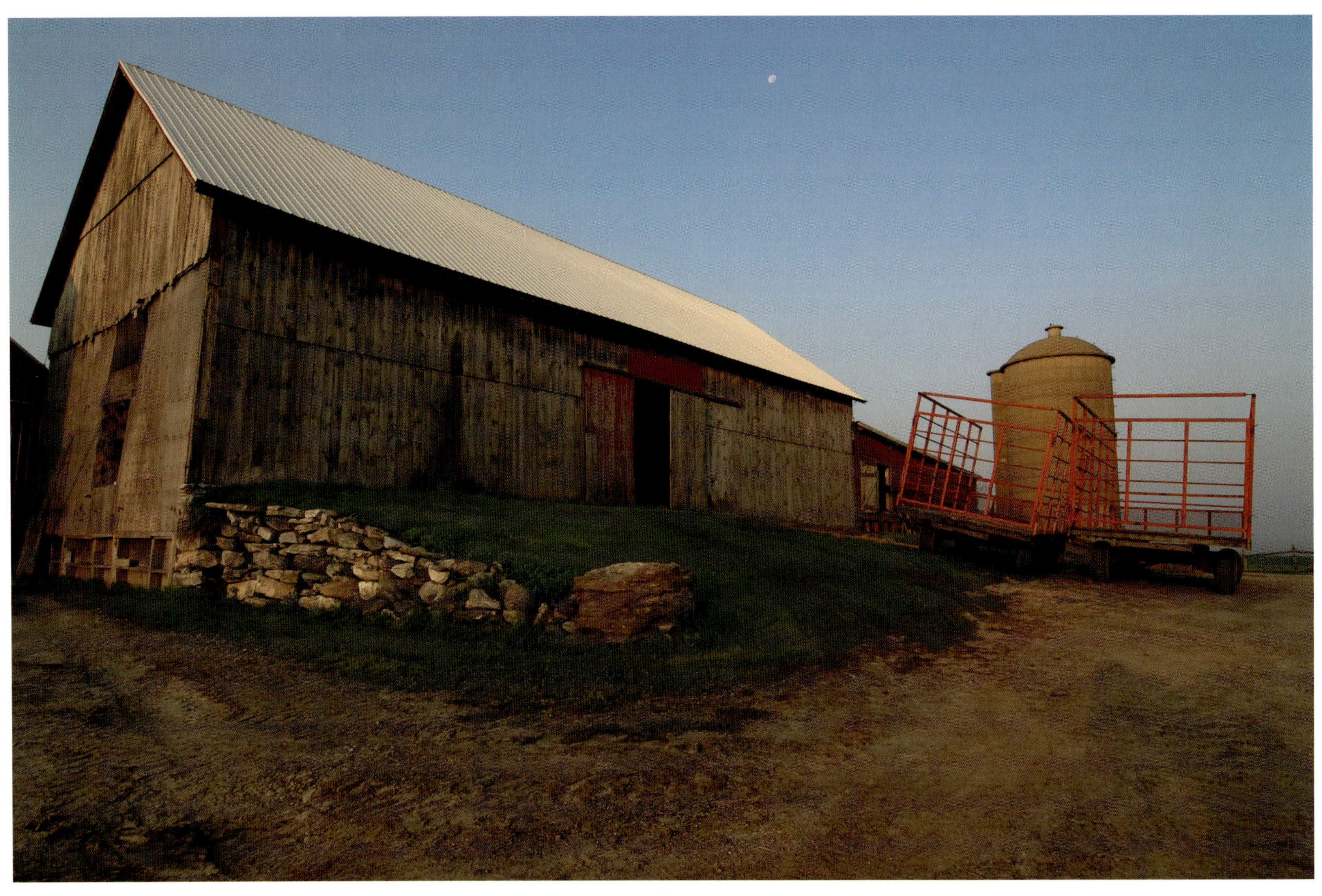

A concrete silo at Proctor's farm on Baldwin Hill, North Egremont.

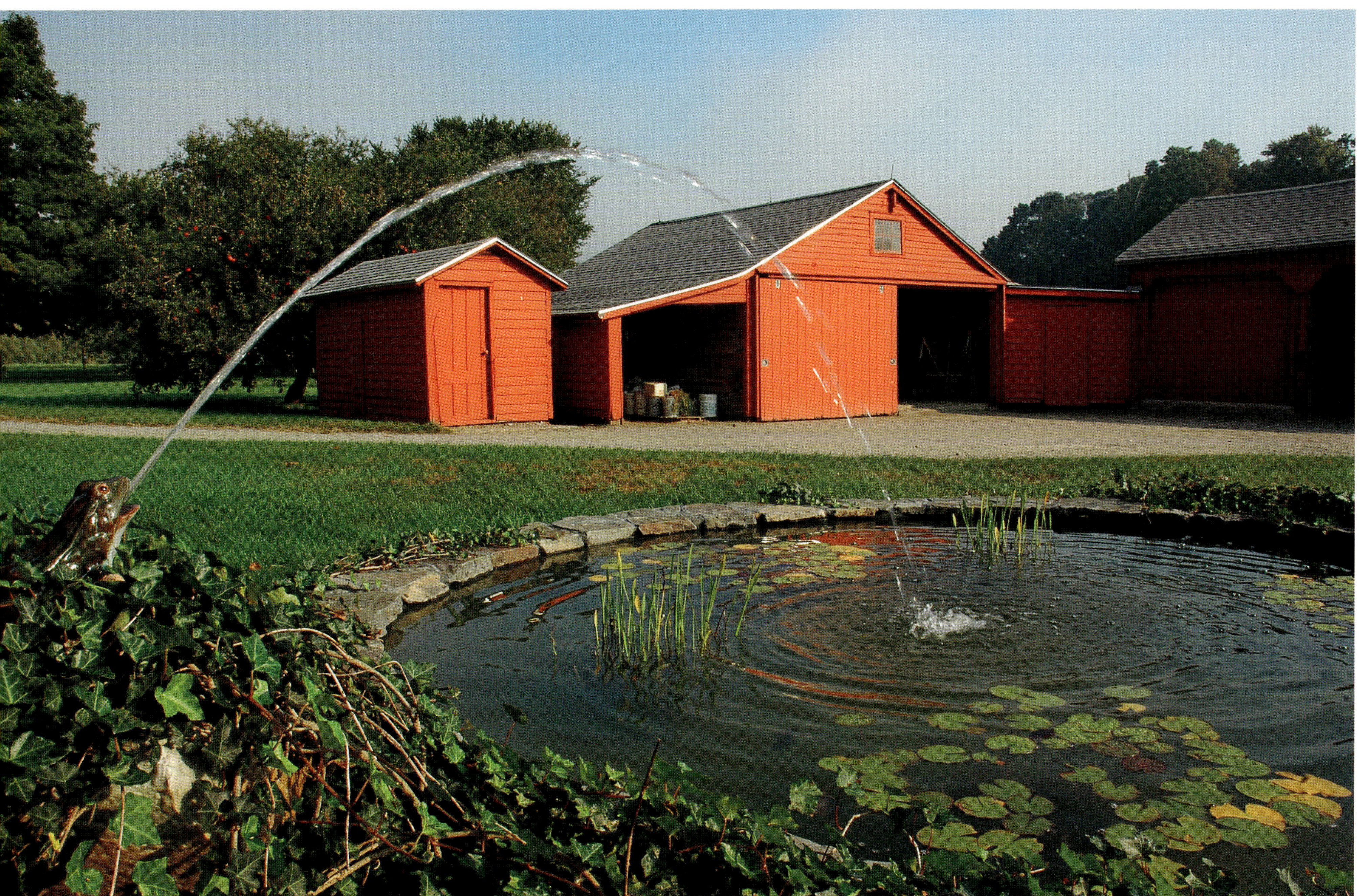

The concrete foundation for the silo at Ron Majdalany's farm. The same year he bought the farm Majdalany sold the silo, along with all of the farm equipment he had acquired. Subsequently he converted the silo base into a fountain.

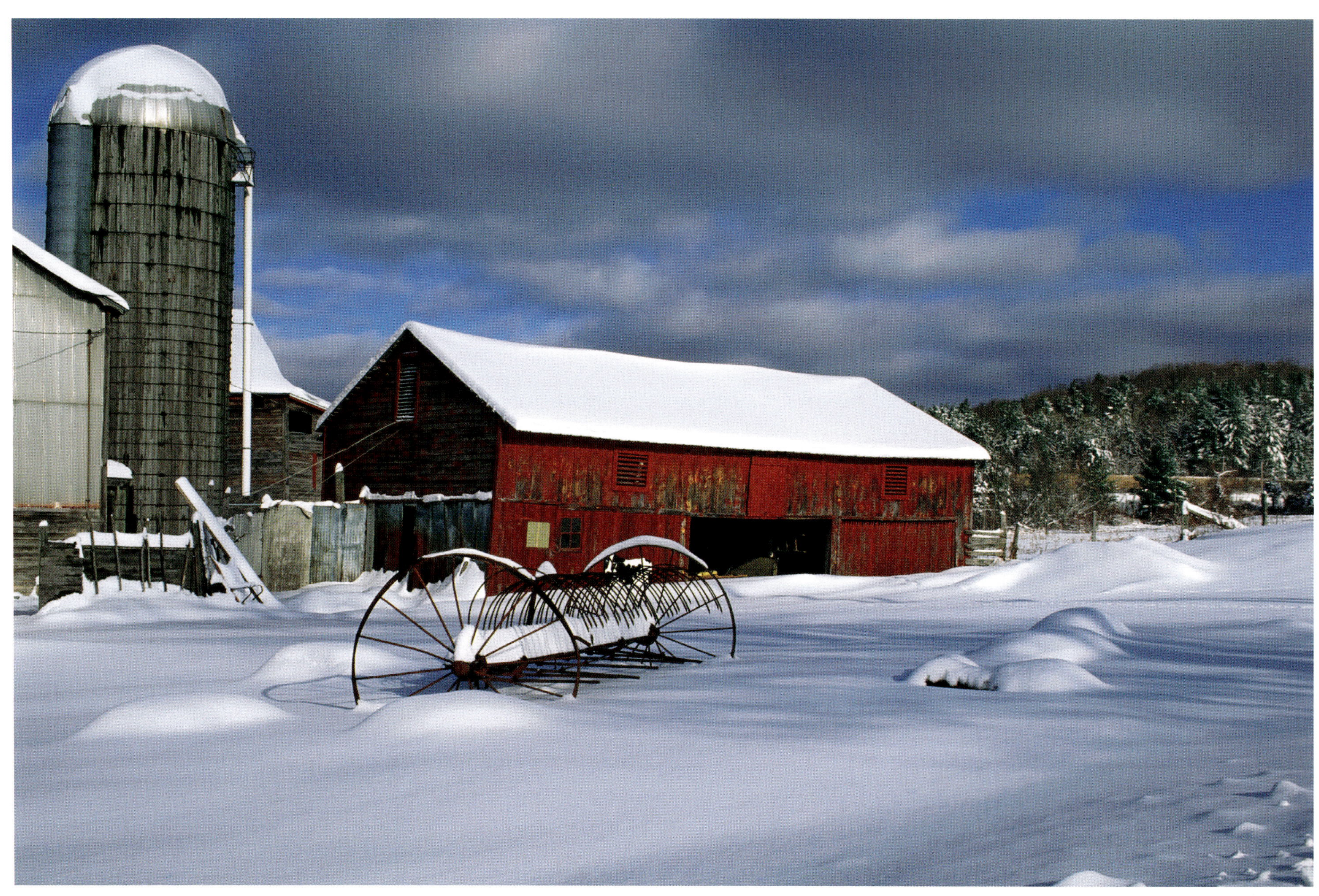

The pole barn at Delmolino's farm, Sheffield. This photograph also appears on the back cover of this book.

The Barns of Berkshire County

Over the last ten years I have photographed many aspects of the landscape of Berkshire County. There are many endearing facets to the scenery here: rolling mountains, valleys blanketed in colorful foliage, dry-stone walls, one-tree hills, and pleasant rivers and lakes, but none are as plentiful and constant as the barns. Although they are a dying breed, there are still many to be enjoyed in their original context, often just off the side of the road. Many, therefore were easy to find and photograph, some had to ferreted out.

There are a number of barns that are located in such fetching scenery, and that possess so much character, that I felt compelled to return time and again to find new ways to capture them through the lens of my camera. I guess you could say that some barns are luckier than others—they were just endowed with more dominant location genes by their creators. But this does not suggest that I assign any more or less value to any particular barn. Nor should the reader assume so just because I have chosen to include one photograph of a particular barn and five of another. Hopefully you will agree with me that some just have more beautiful sight lines and unique characteristics than others. Sometimes one picture is all that is necessary to show how beautiful a particular barn is. And, truth be told, sometimes, despite returning to a barn more than once, there was just one image I got that showed it in all of its possible beauty. Serendipity plays a role in the process, too, and often weather alternates between being my best friend and my worst enemy. Sometimes I get just one opportunity for a great shot.

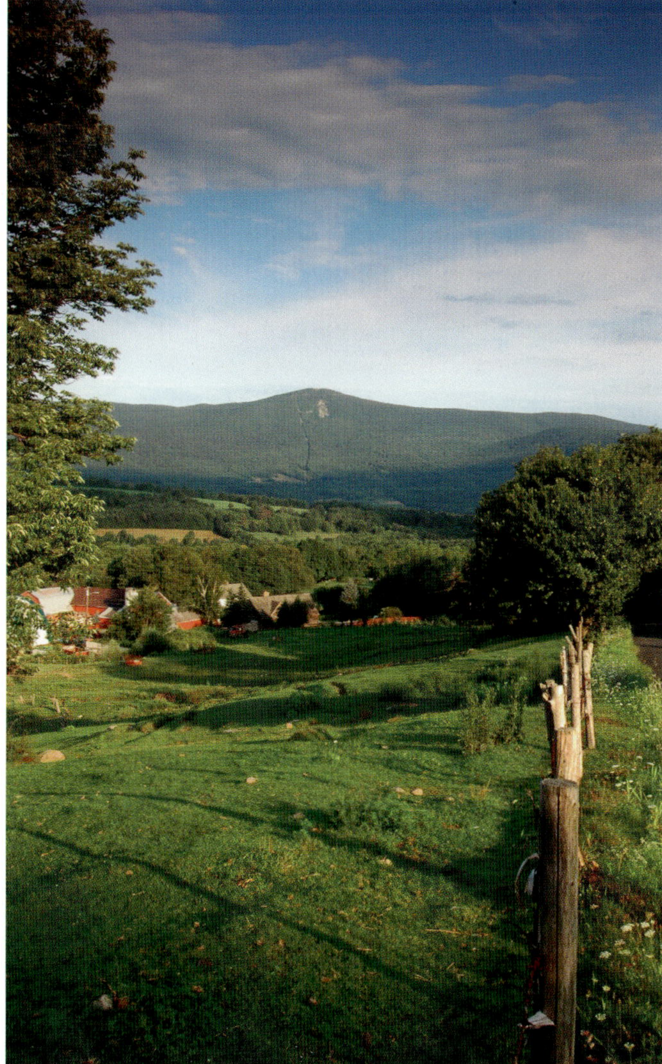

Further up the hill from the Alibozek Family Farm barn complex that is visible in the center left of this photograph you have a beautiful view of Mount Greylock. This area is called Little Egypt.

Whatever the case, I have tried to compile a book of photographs that best represents the population of barns in Berkshire County, and to give all parts of the county at least some coverage. When it comes to barns there is beauty all around in the Berkshires, but it helps to have time on your hands and the combined knack, inclination and enjoyment of exploring back roads and, occasionally, getting lost. I have not provided a map, just the general locations of the barns—you must BYOM (bring your own map). If this book is your only journey of discovery of the barns of Berkshire County, I hope that you enjoy it as much as I enjoyed the ten years of work that it represents.

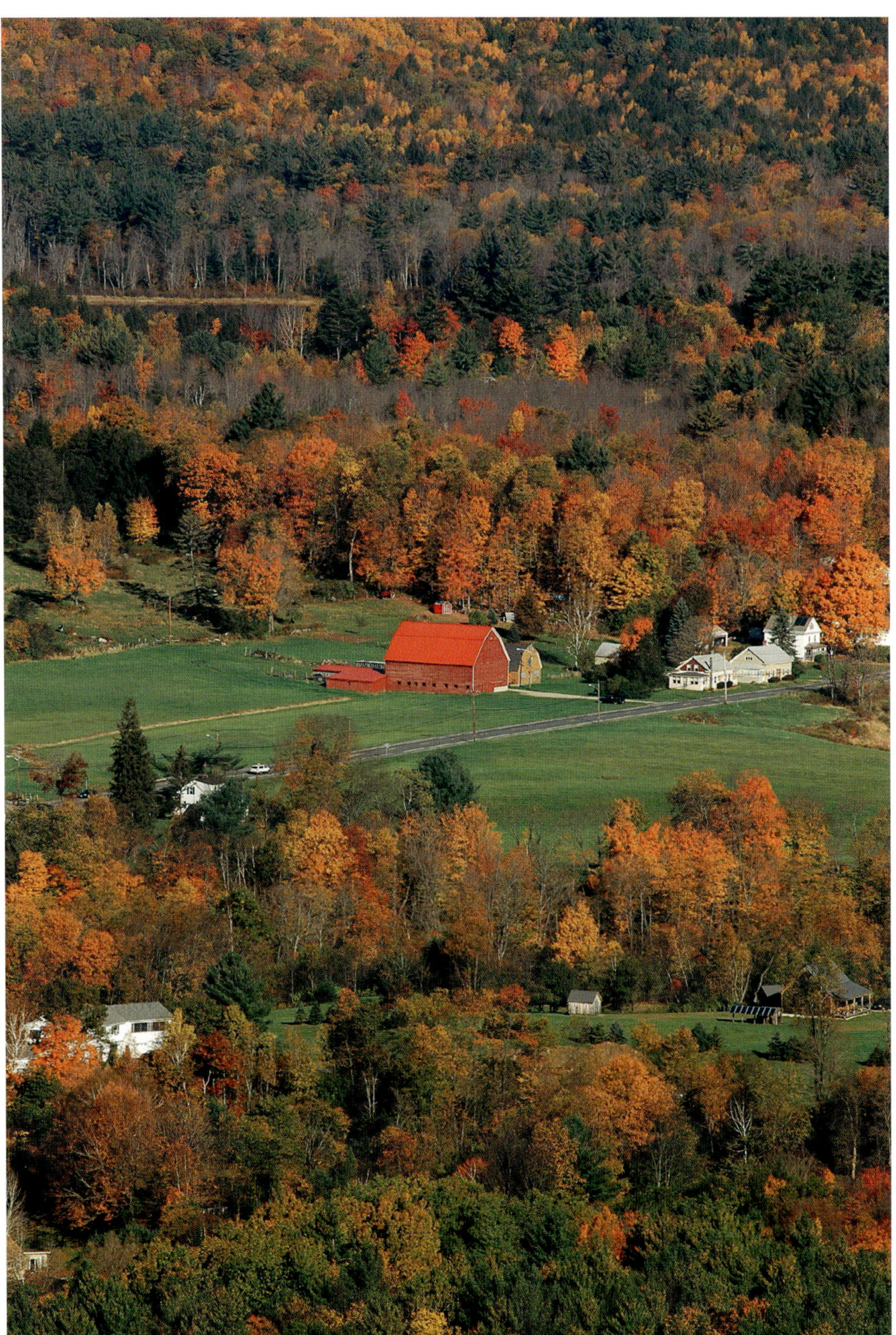

The red barn on Route 8 in Clarksburg must catch every driver's eye as they descend Route 2 into North Adams.

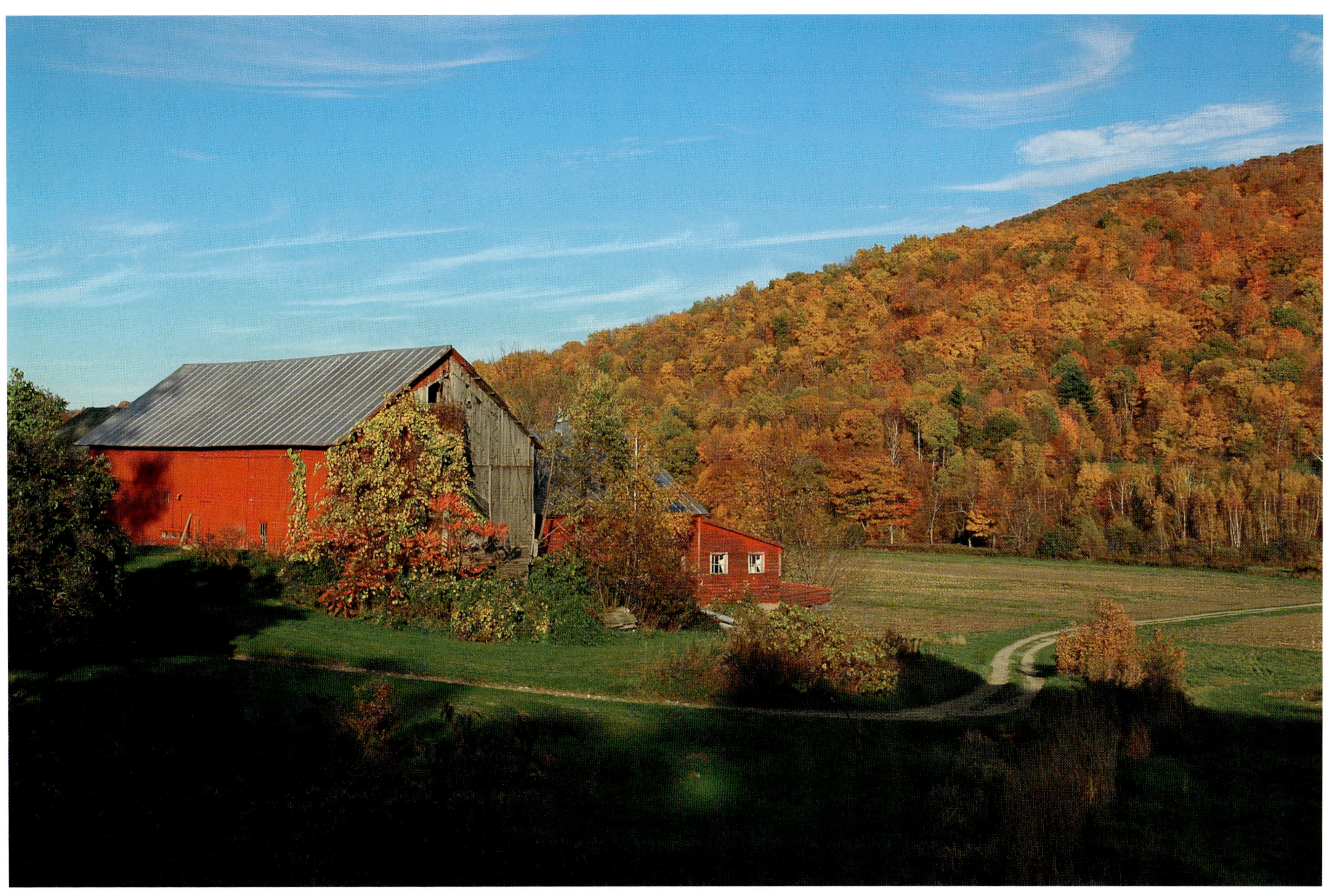

This is an example of a barn that I happened across when I was on my way to a specific location to photograph fall foliage. I stopped for no more than five minutes, took several shots, and quickly moved on. This barn is on Route 43 in Hancock.

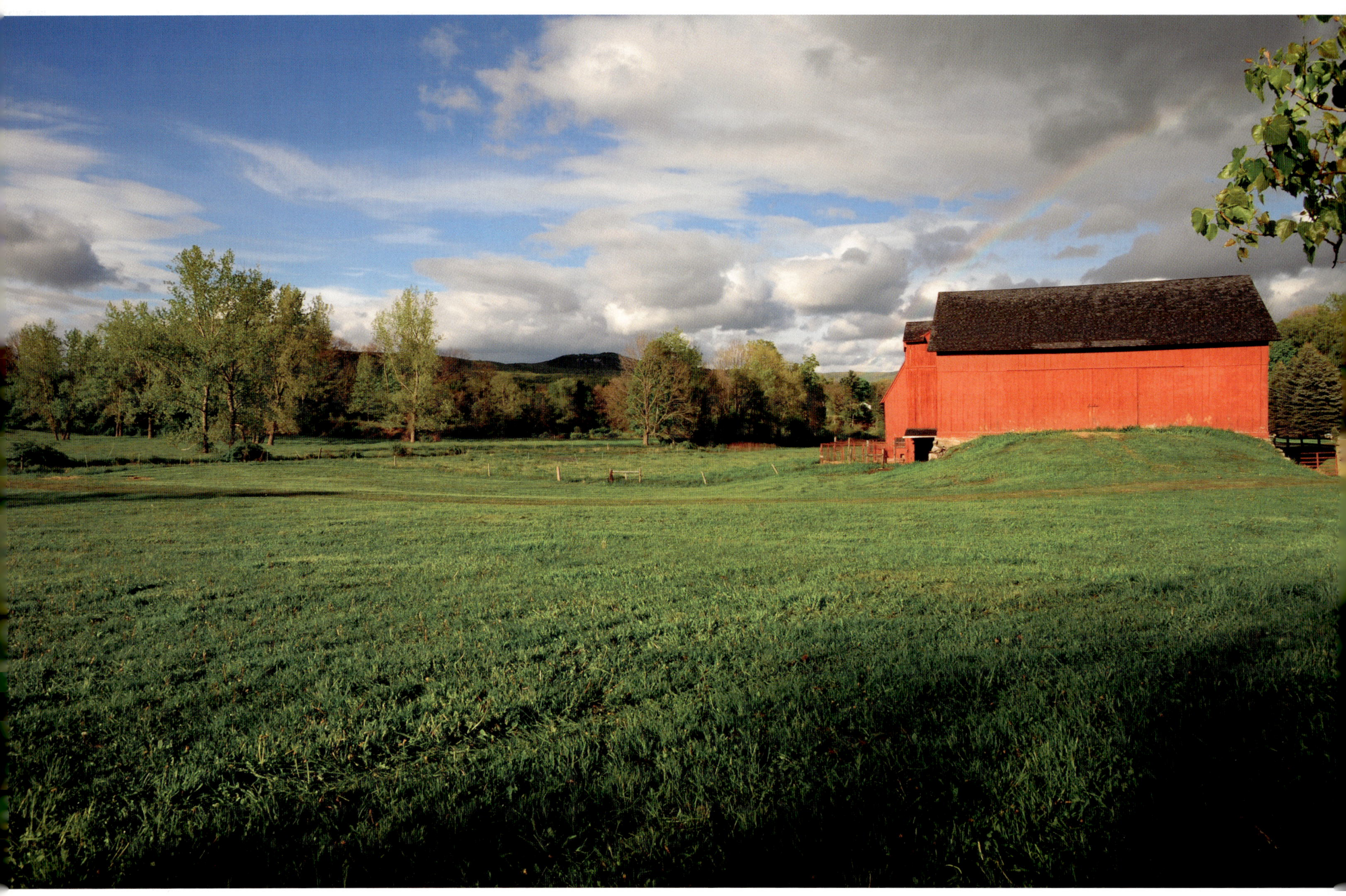

Monument Mountain is seen in the background of this view of the 1875 bank barn on Division Street where a clearing storm produced beautiful lighting and a rainbow over the barn.

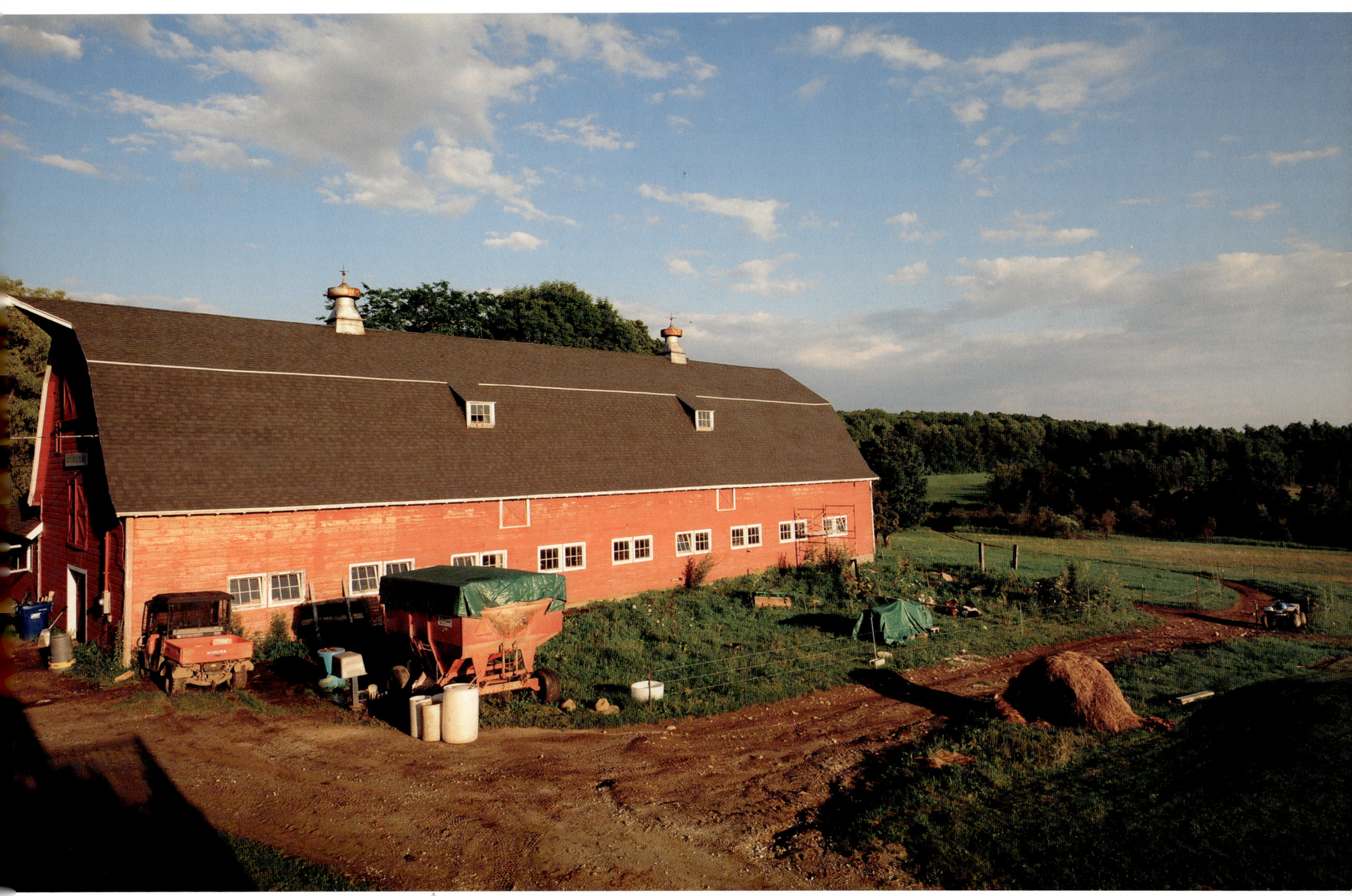

This gambrel roof barn has a spectacular setting at the top of Blue Hill Road in Monterey. Based on the frame construction of the roof, this barn was probably built in the early 20th century.

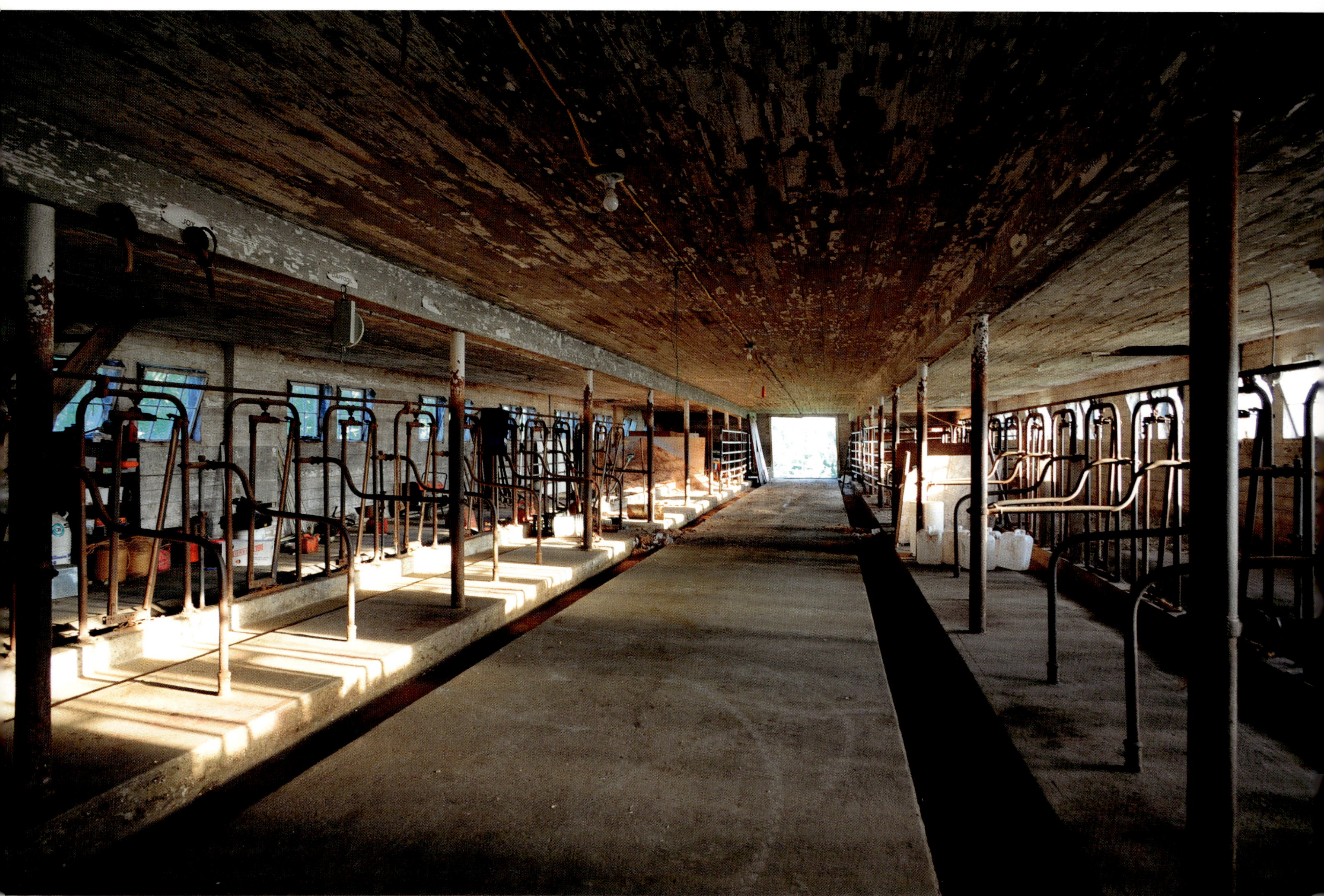

The milking floor at Blue Hill Farm.

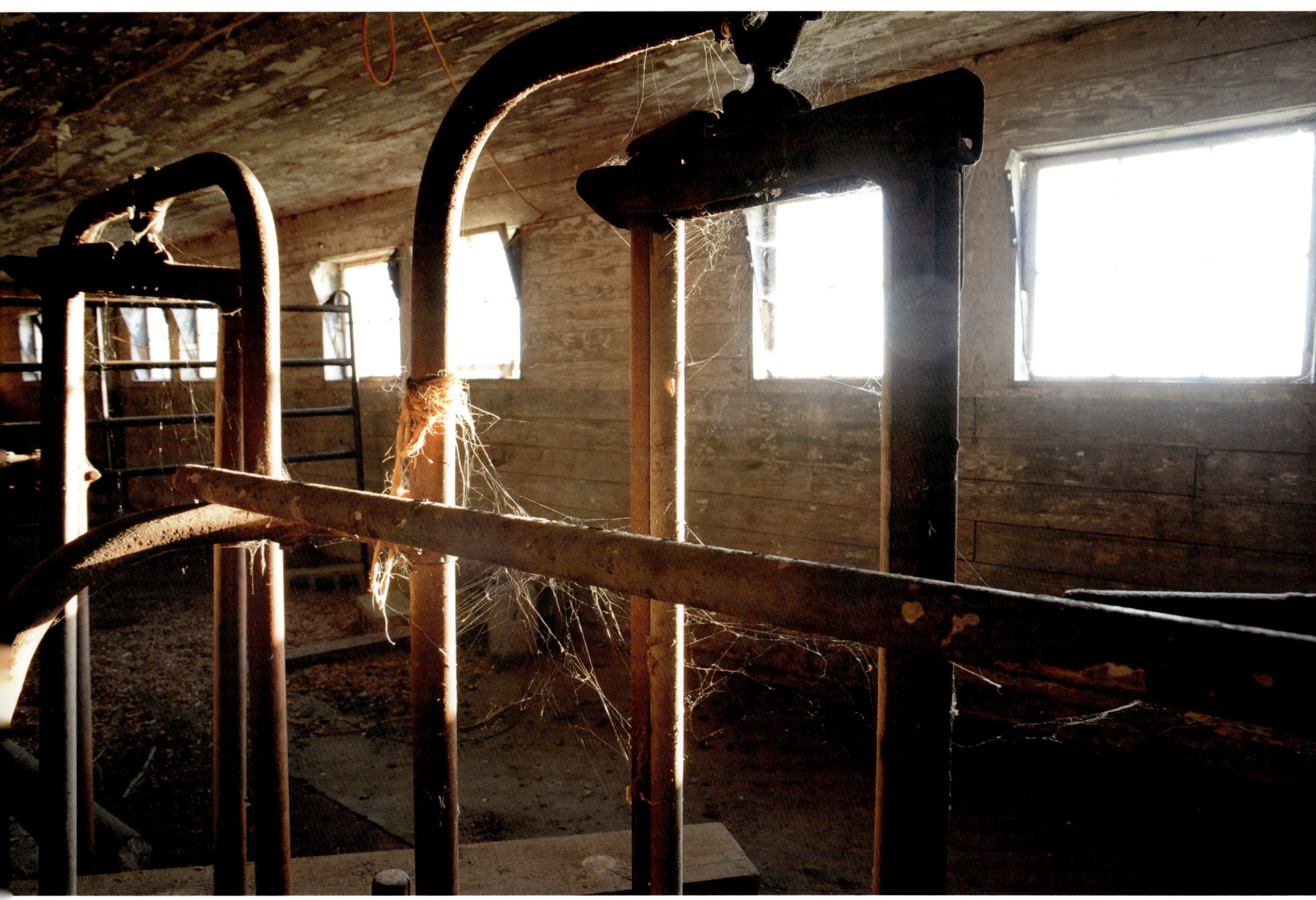

Light streams in through side windows at Blue Hill Farm highlighting cobwebs in the stalls.

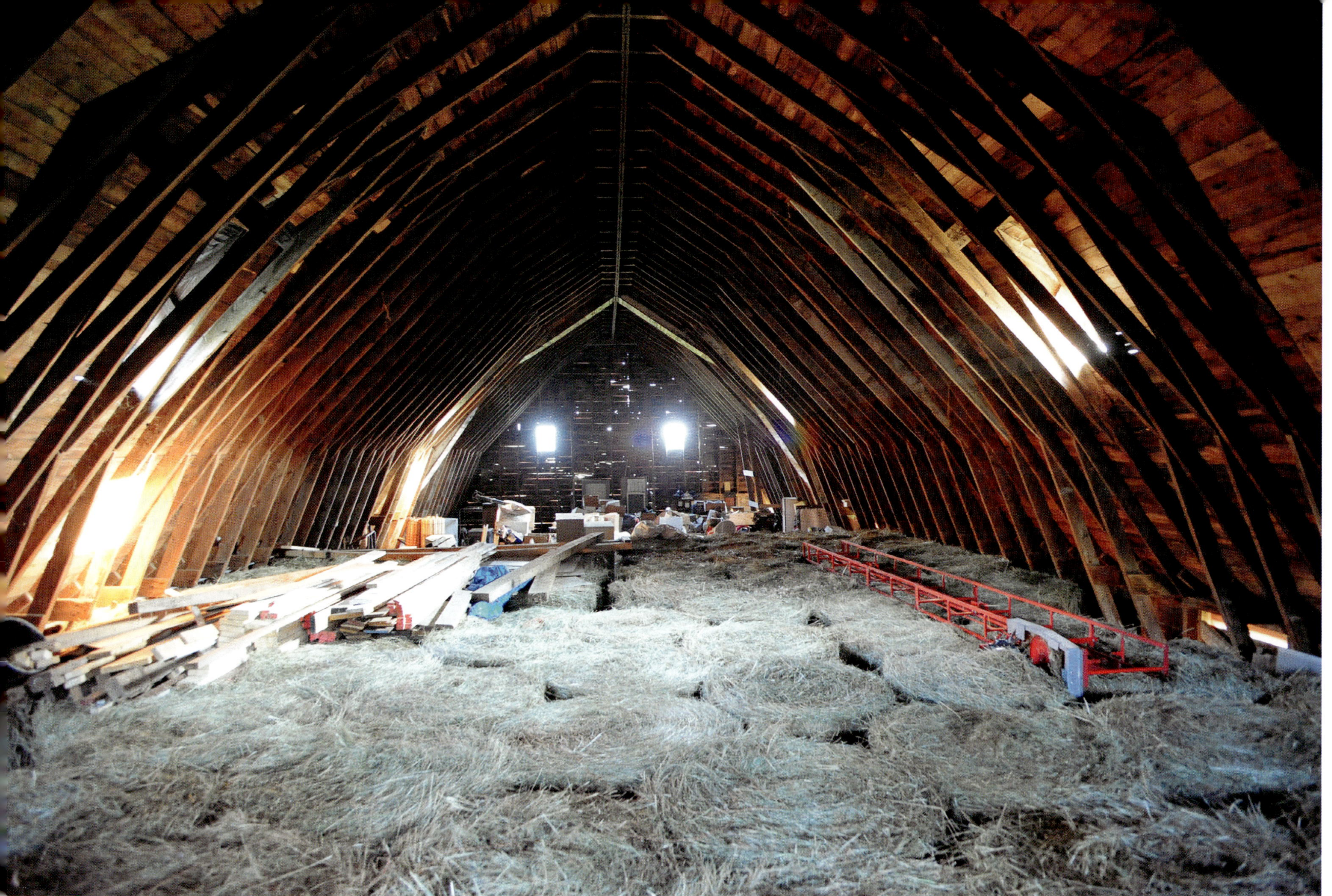

The roof structure of Blue Hill Farm is a good example of building design techniques that were made possible by new technologies that were rapidly surfacing and evolving through the 19th century. The design of this clear-span, gothic-arch-truss roof was made possible by the introduction of smaller timbers and newer framing concepts that abandoned the post-and-beam principles. This is an example of what became known as balloon framing.

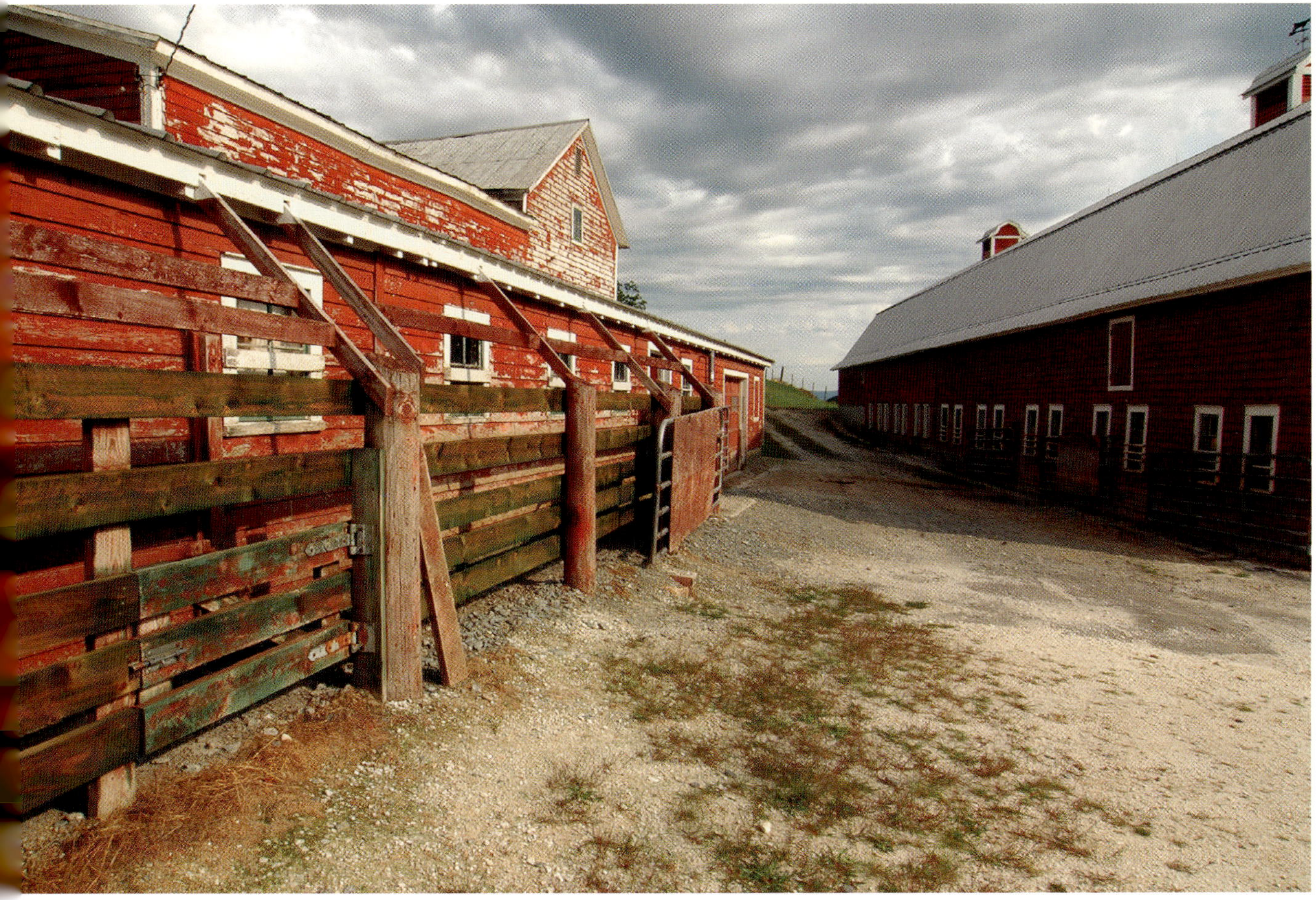

According to owner Ed Clairmont, the oldest barn at Gulf Farm, seen here on the left, dates to the 1800s. The larger gambrel roofed main barn that displays the Farm's moniker at its gable end, was built in two halves, with the first part completed in 1936 and the second half in 1972.

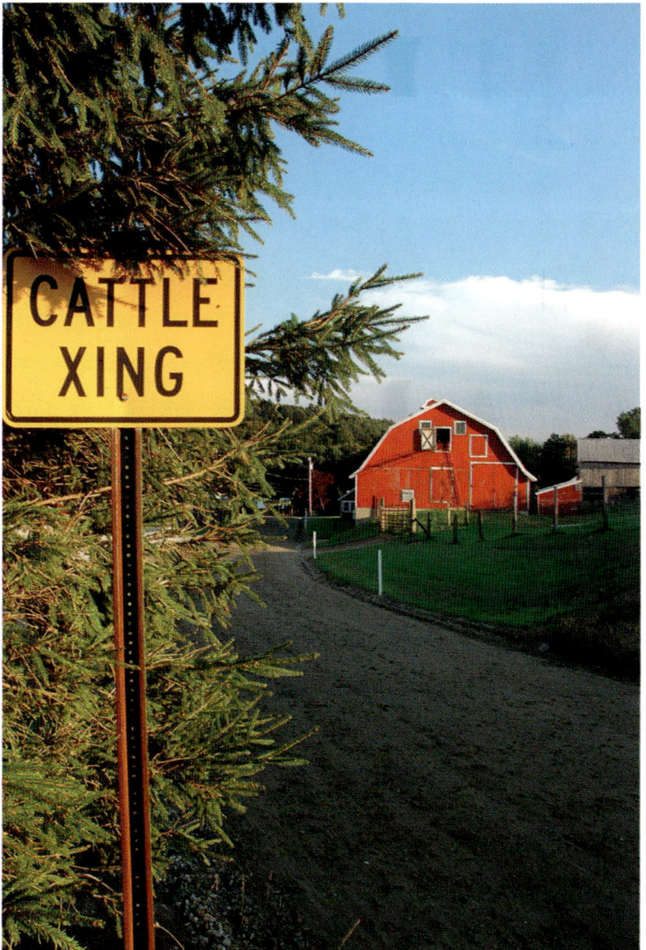

The north end of the main barn at Gulf Farm seen from Stewart White Road.

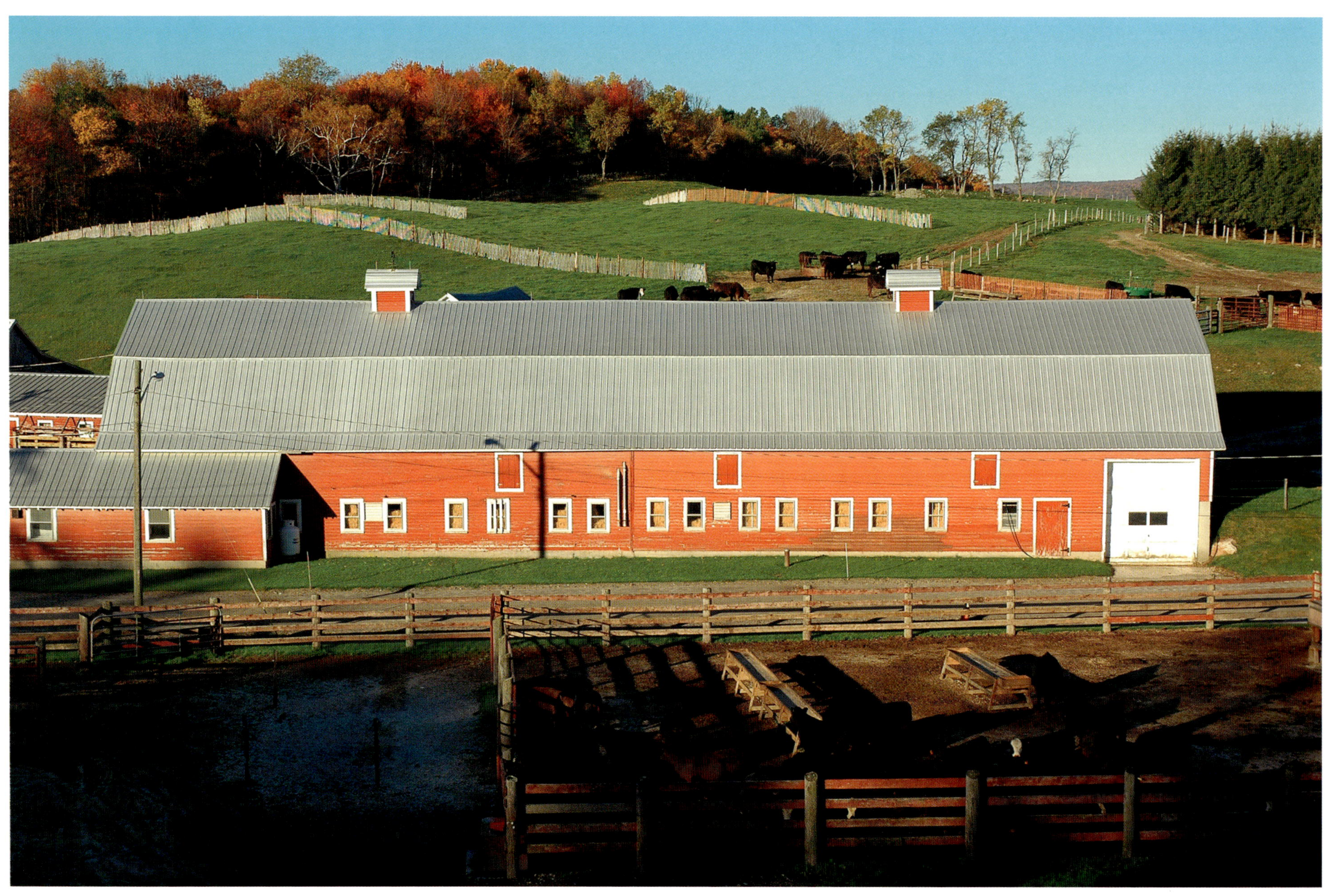

The main barn at Gulf Farm seen from the eastern hillside across Stewart White Road.

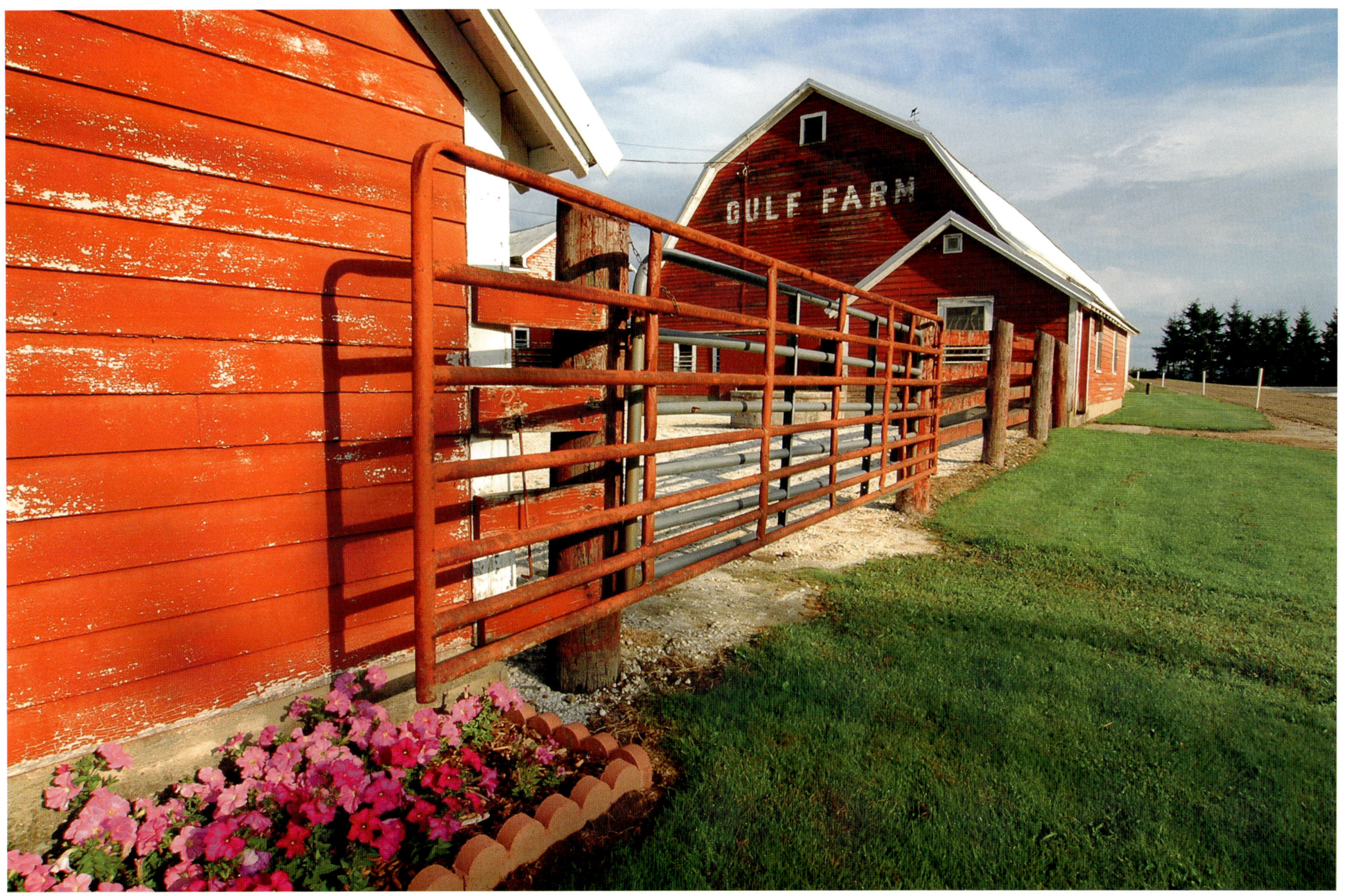

The southern-facing gable end of the main barn at Gulf Farm with the farm's name prominently painted on.

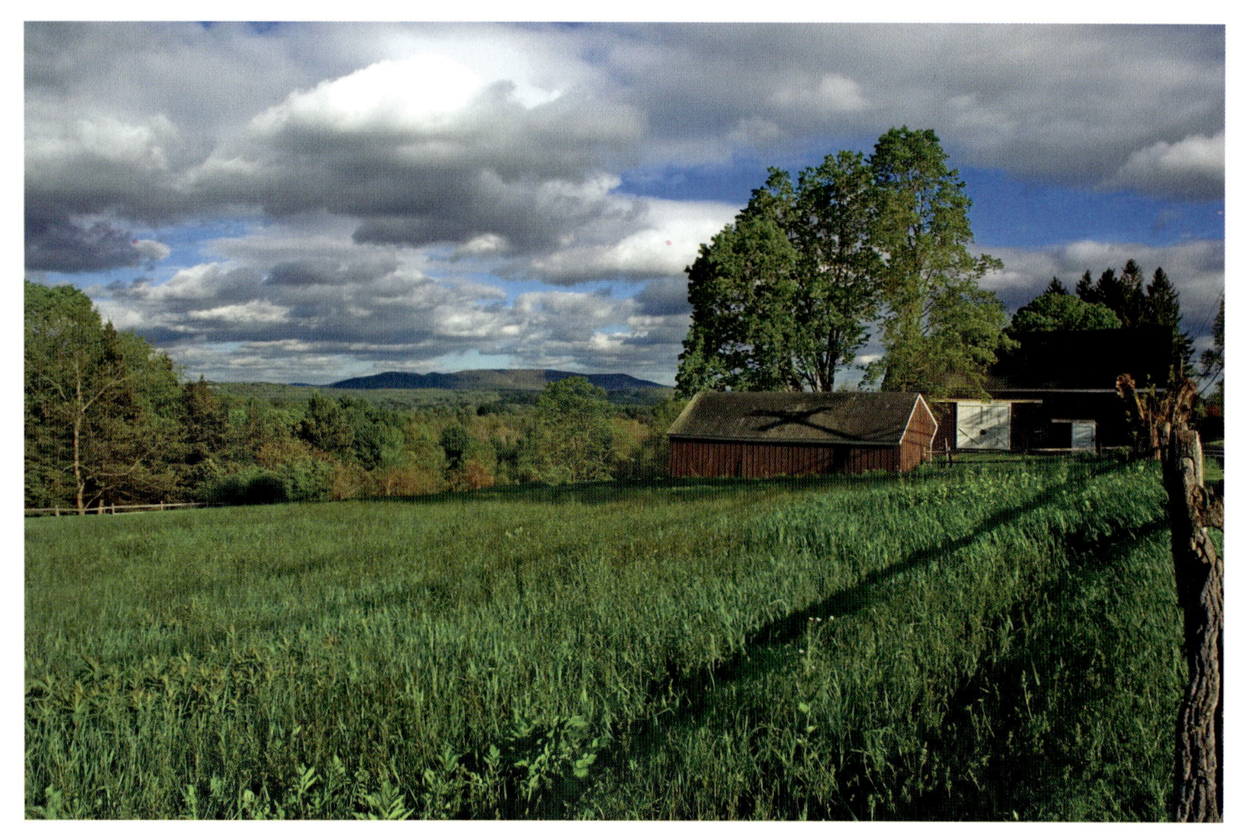

Seekonk Cross Road in Alford looking southeast toward Great Barrington and East Mountain. I have no information about the date or origin of this barn.

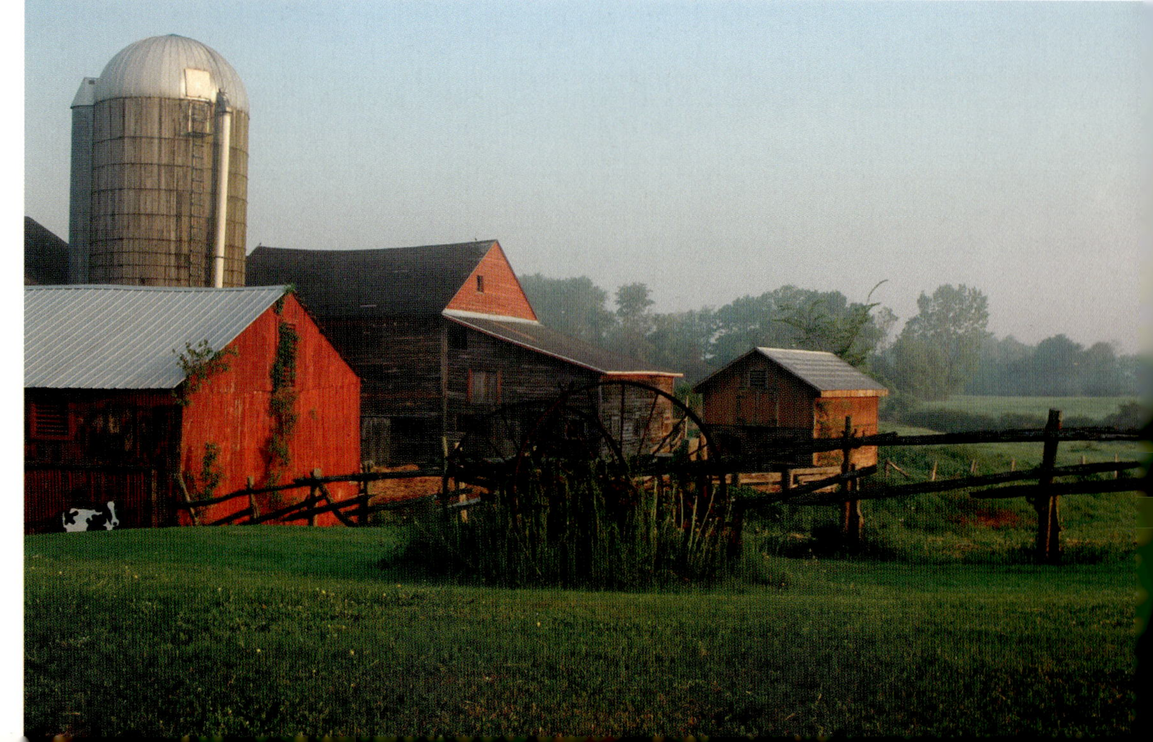

Sunrise at Delmolino's farm, Sheffield.

Winter scene at Delmolino's farm taken from the southwest on Sheffield-Egremont Road.

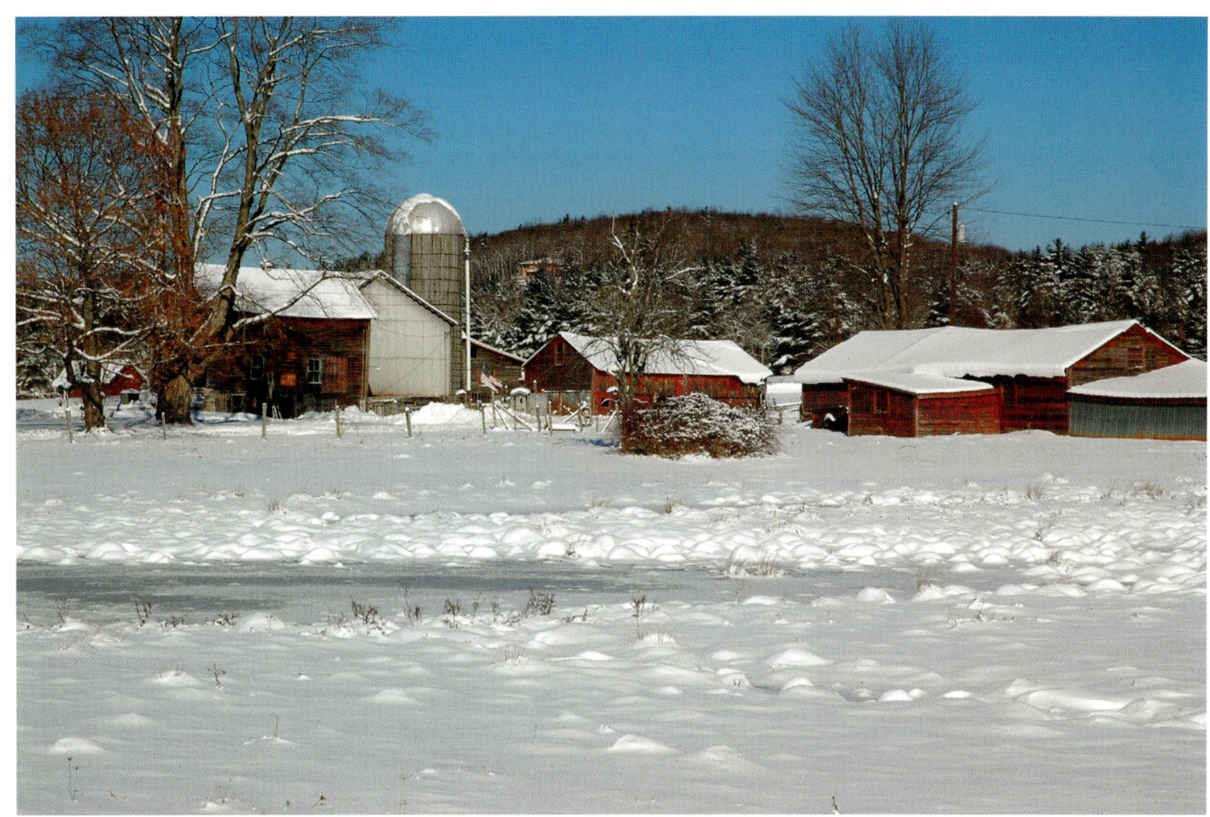

Clarksburg native Ken Demers has spent all of his 80-plus years here on the farm that was built adjacent to his ancestral home, which he says is the oldest house in town. He does not have a construction date for his old hay barn but believes it is probably of English origin. His horse Johnny stands sentry at the gate.

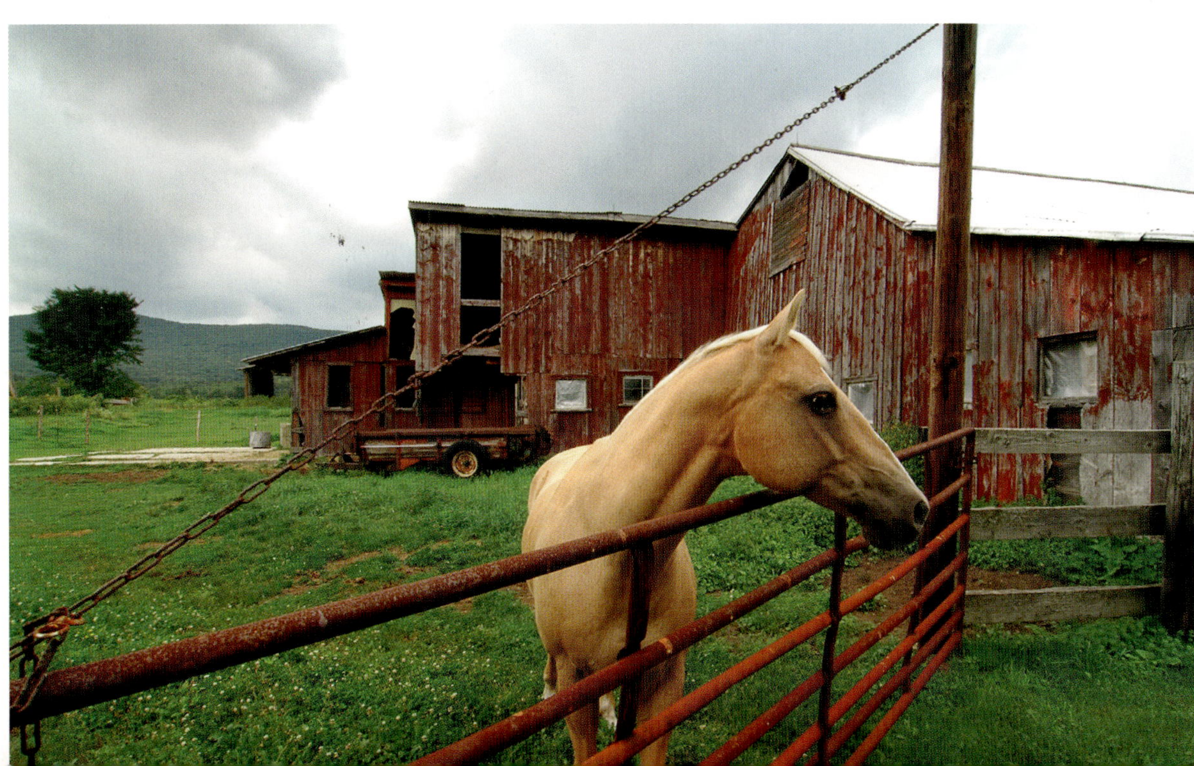

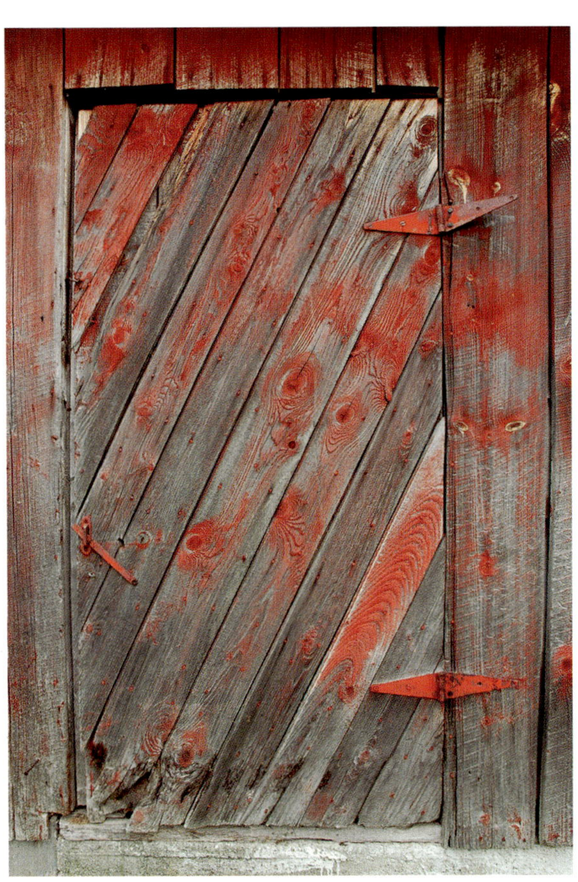

A weathered door on Ken Demers' barn.

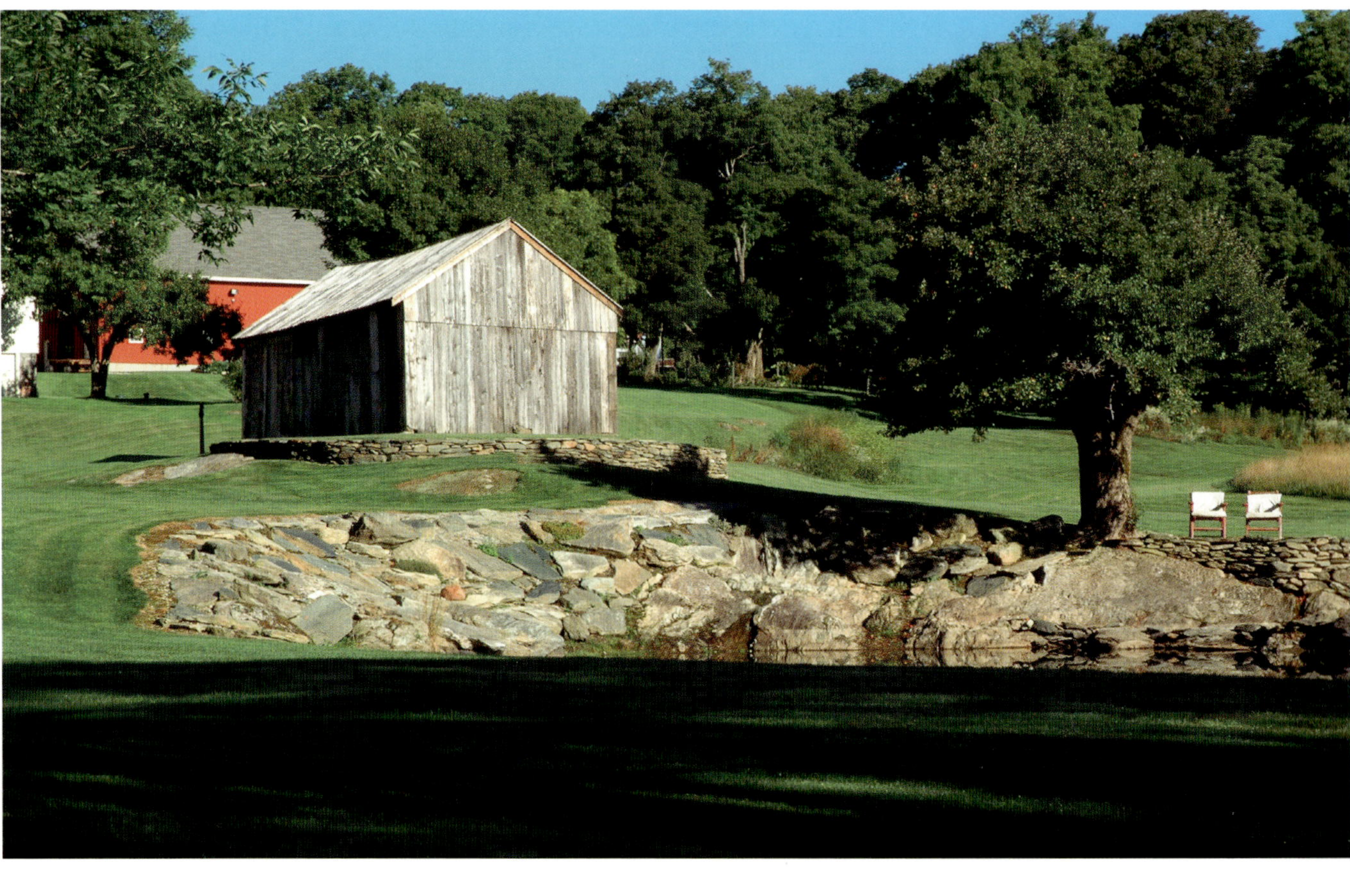

This building was actually a sugar house and it is a popular icon for all who enter Berkshire County in Savoy on Route 116. According to owner Bob Veronesi, it was built with a pegged post-and-beam frame on the original Remmington homestead. Entries in the family diary make mention of sap being boiled in the shed in 1881, so the barn is known to be at least 125 years old—the Veronesi's believe it could be as much as 150 years old. Having been battered by a severe storm in the fall of 2005, the building was in such a disintegrated state that Veronesi, after taking some drastic steps to try to save it, was at his wits end and was ready to raze it. Just hours before he was set to tear it down the barn got a stay of execution due entirely to a photograph an admirer had taken of it. This is the building, fully restored, in August 2007.

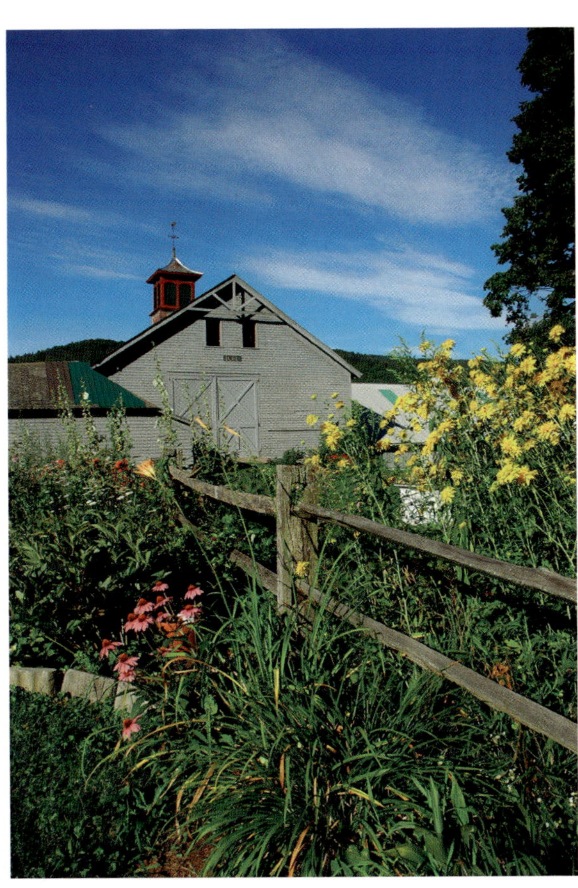

Although Sunset Farm was originally established in 1704, this dairy barn wasn't built until 1891. Until 1862 the farm had changed hands four times. It has been owned continuously by seven generations of the Hale family since then. Frustrated by the chronic financial pressures of a working farm, the Hales sold off the dairy herd in 1991 and converted the farm to a bed and breakfast that has become a popular venue for weddings and special functions. The dairy barn now serves as a sort of great hall for these occasions.

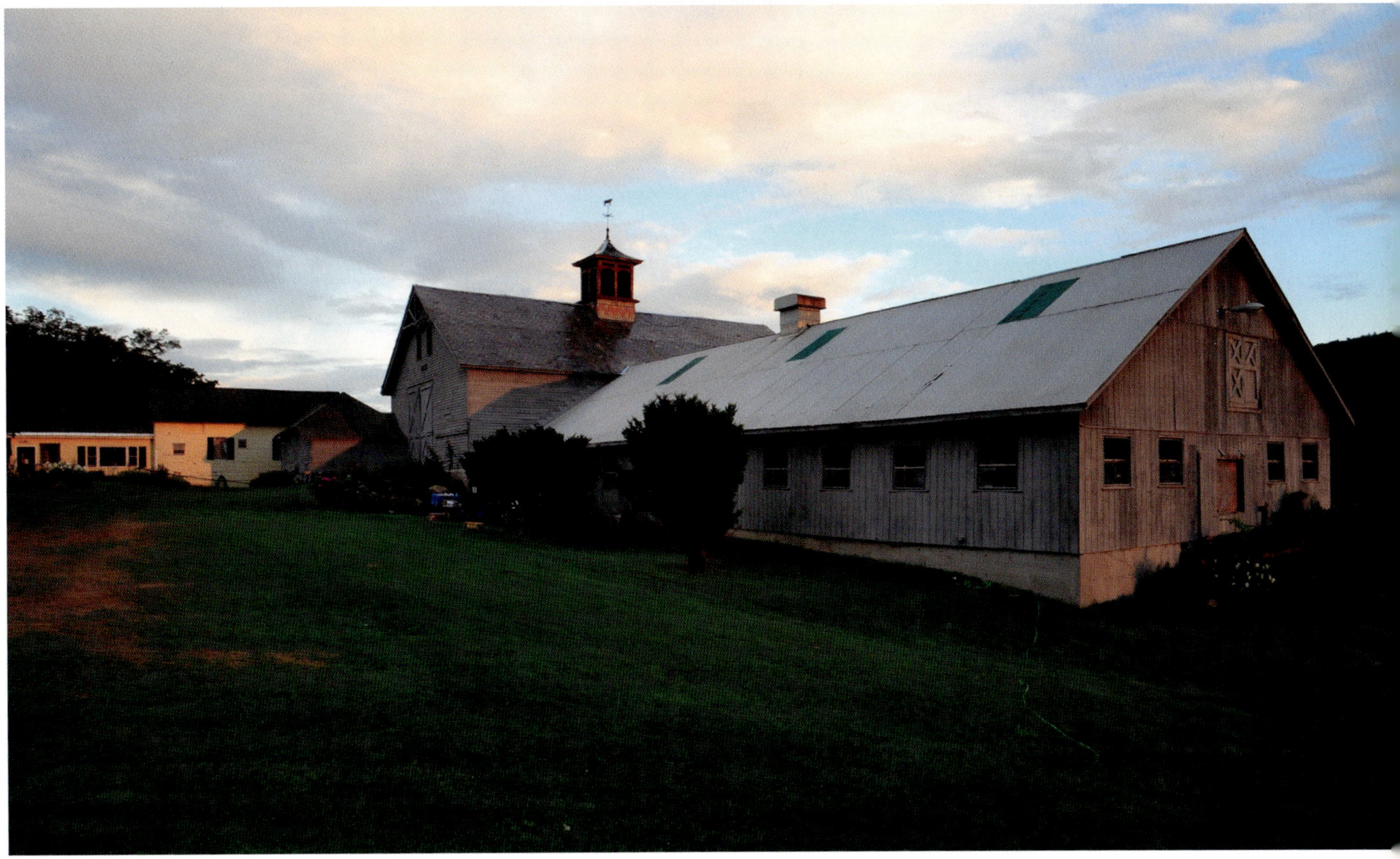

Sunset Farm, which is perched on a hilltop at the end of Tyringham valley, is aptly named since it has an unobstructed view to the western horizon. In this shot the farm basks in the final soft glow of another beautiful sunset in August 2008.

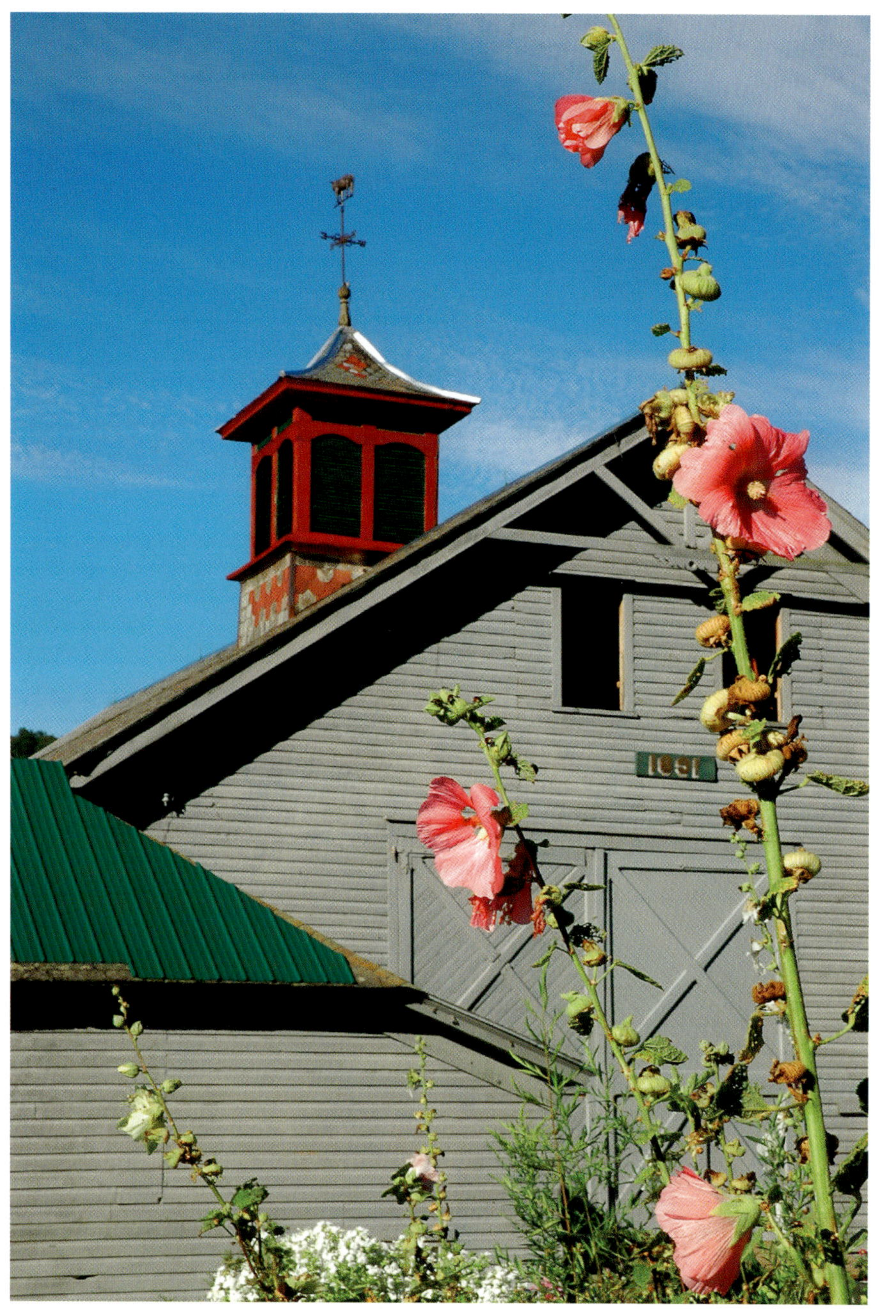

The ornate cupola and decorated gable end on the dairy barn at Sunset Farm provide an example of how barns were becoming more fashionable at the end of the nineteenth century.

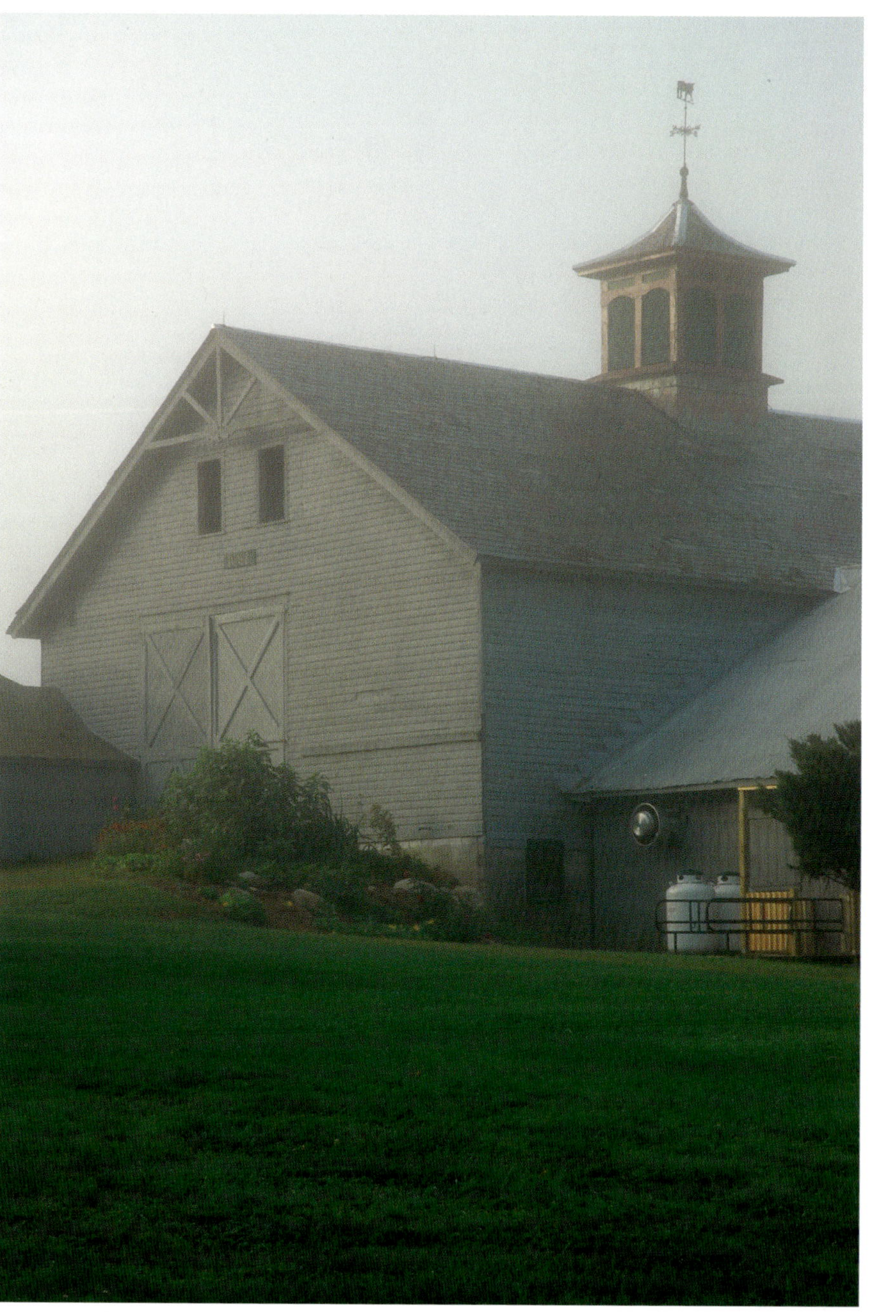

A misty sunrise at Sunset Farm.

The gambrel roof bank barn (shown earlier), and the gable-roof side hill barn on the right, are located on the east side of Route 41 in Sheffield. The gable-roof barn is probably the older of the two, and both would most likely date to somewhere between 1840 and 1890.

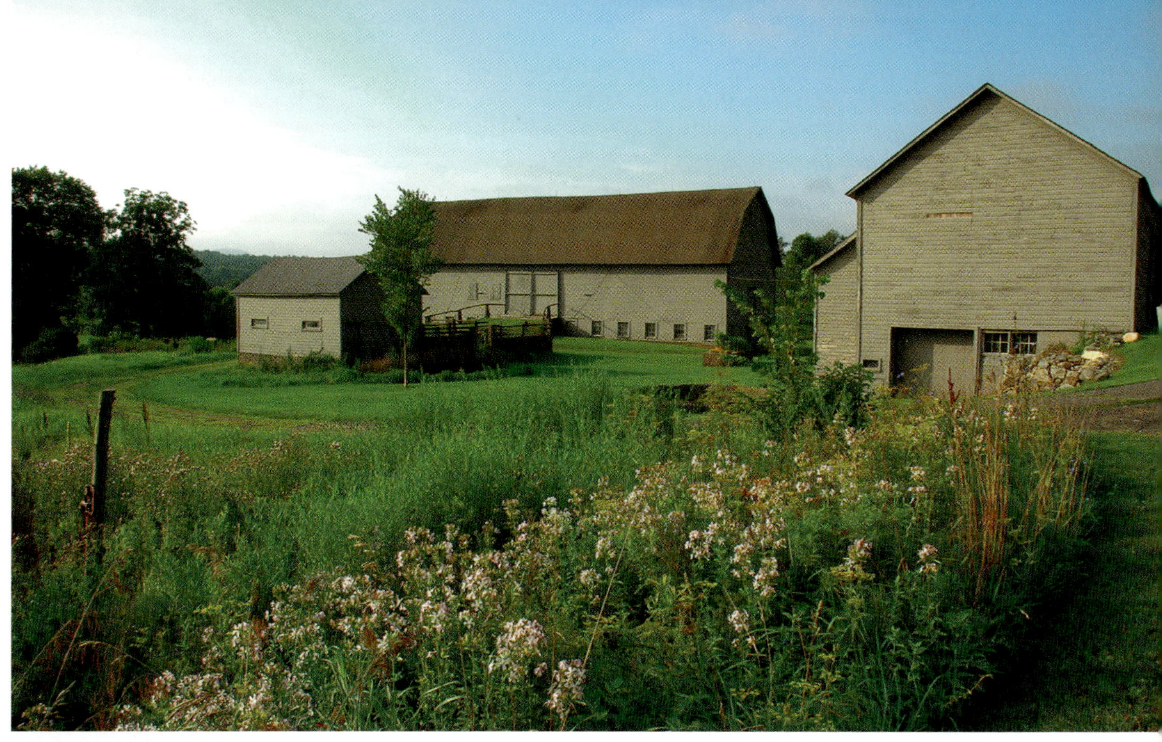

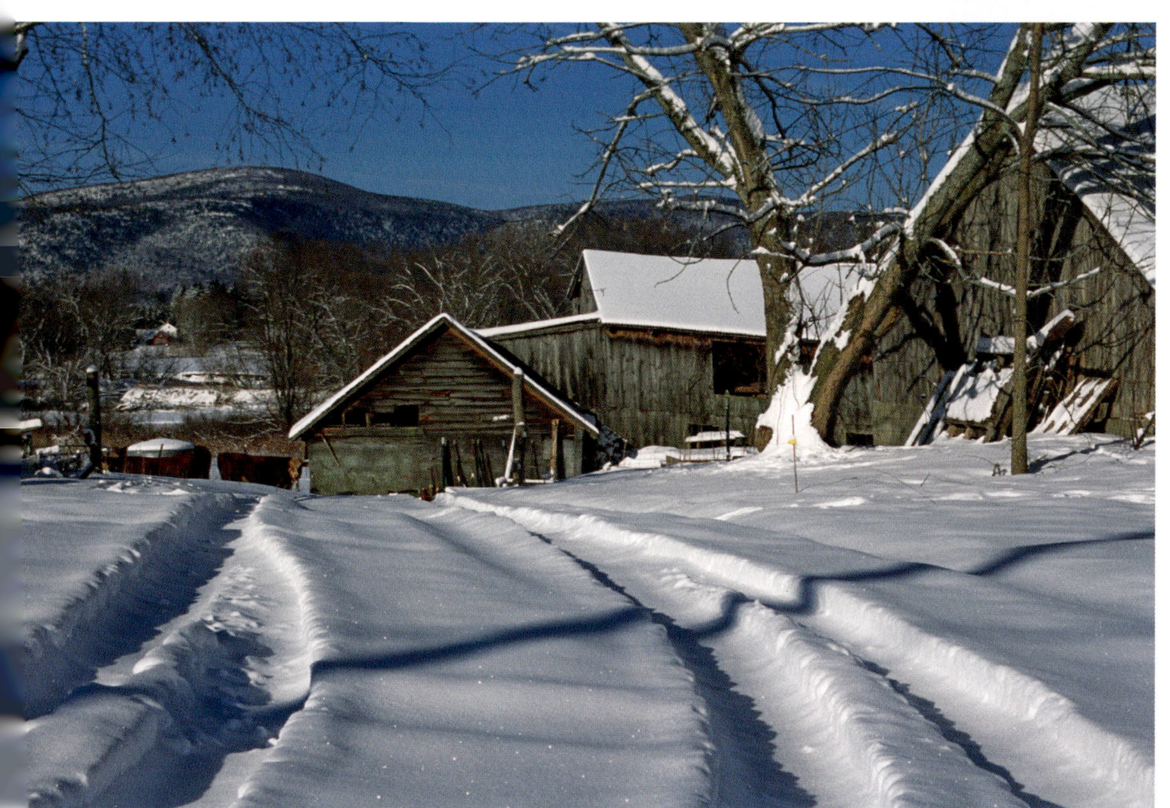

This barn on Boardman Road in Sheffield no longer stands.

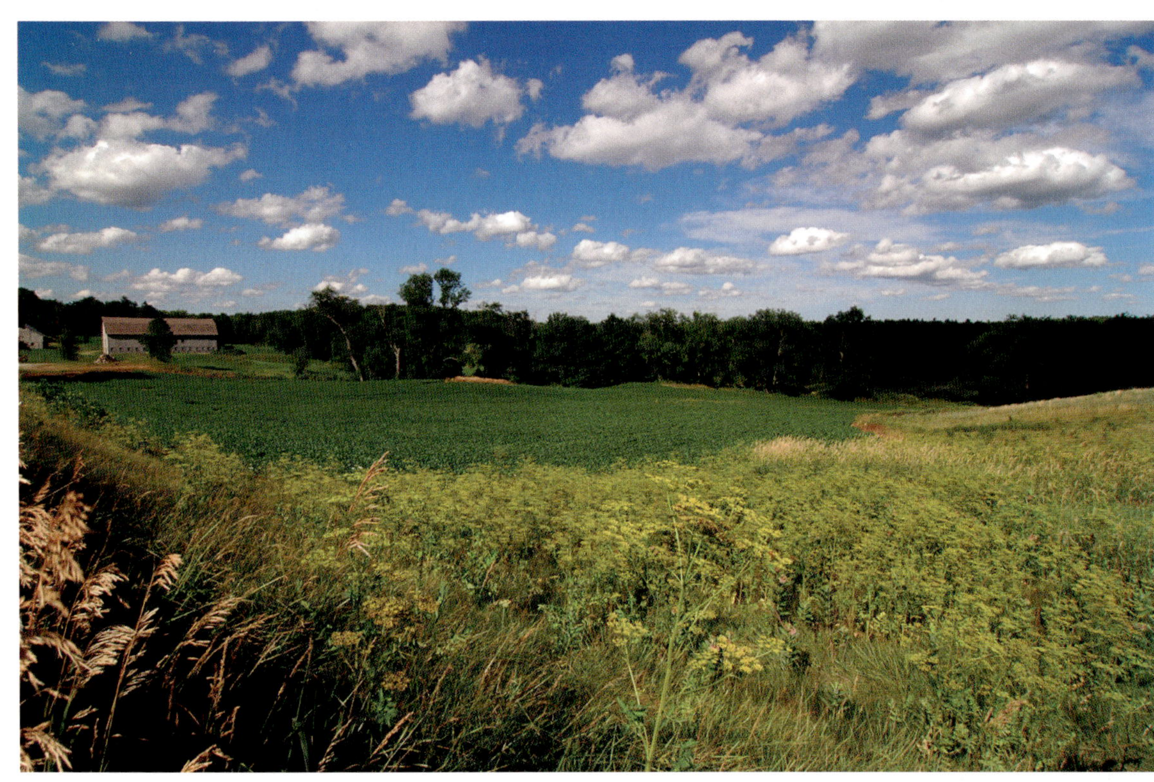

Sometimes, even though it is not the prominent feature in a landscape, a barn will still add the essential finishing touch, as seen here where the two barns on Route 41 appear in the far left of the photograph.

While on the hunt for shots of fall foliage in the Williamstown area I happened across this scene of a barn that appeared to be peering around the corner. I set up my tripod, took a few shots, and moved on. That was my only encounter with this barn.

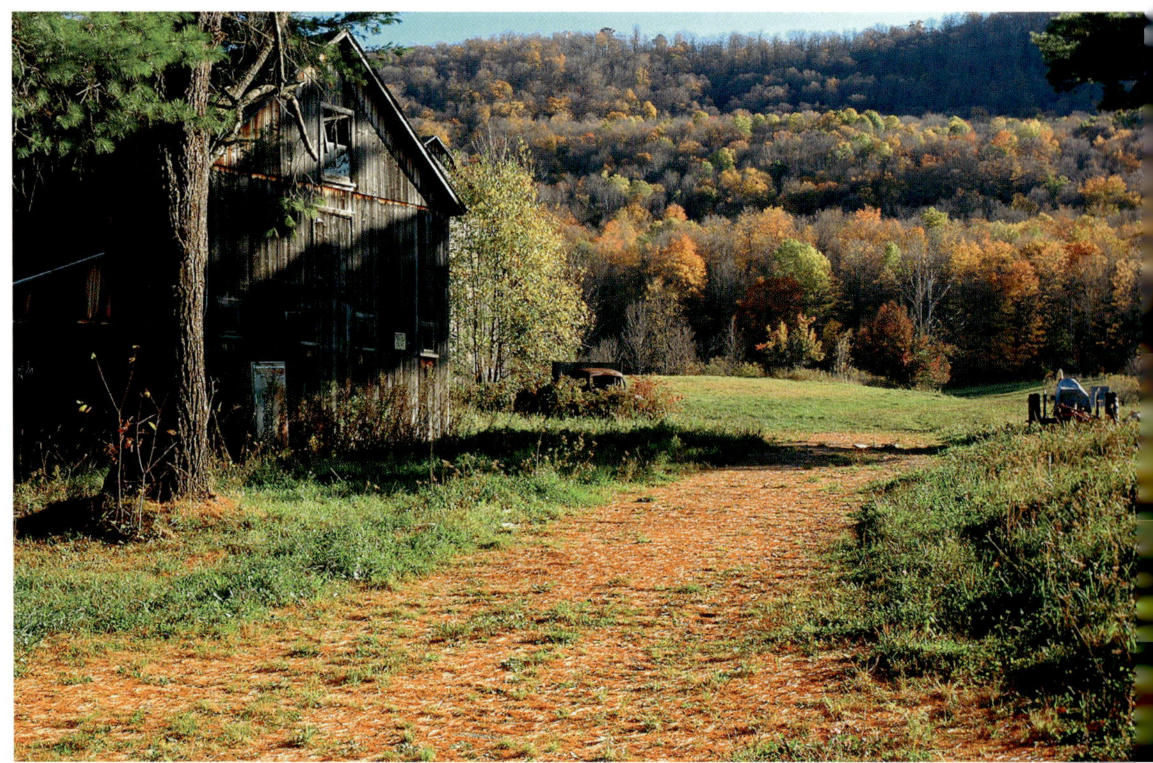

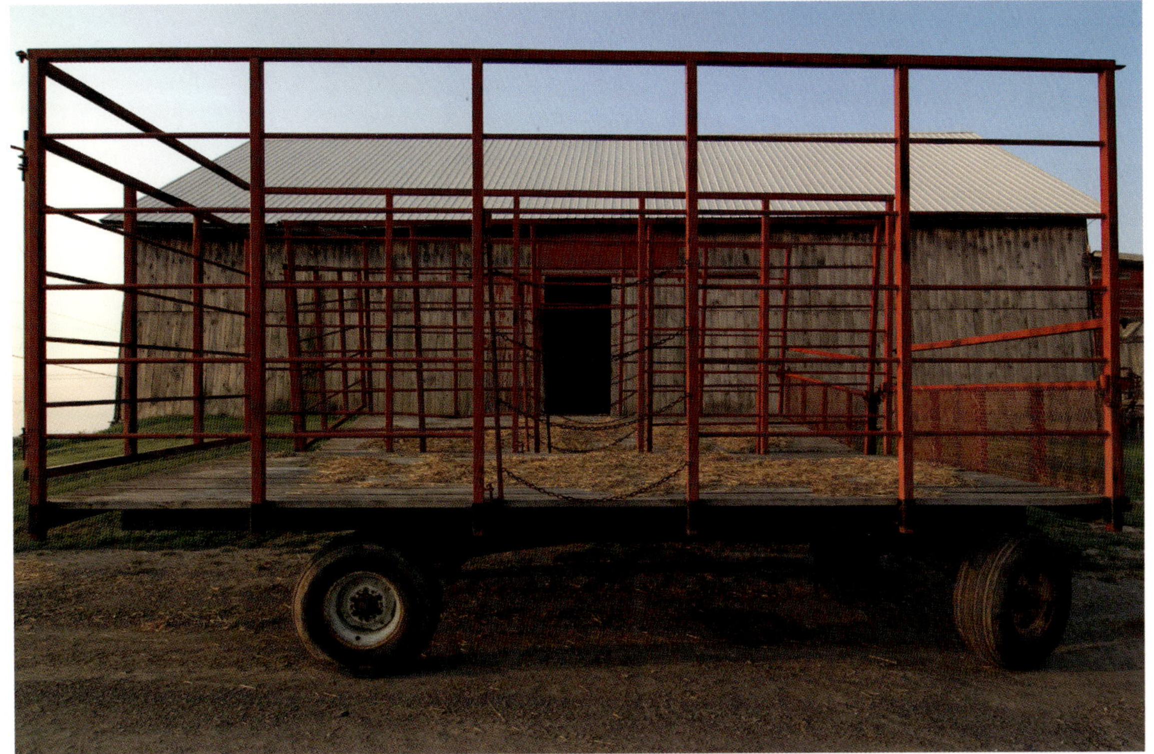

The aligned geometry of two hay carts creates an unexpected vanishing point at the open doorway of an old side-entry barn on Proctor's farm in North Egremont.

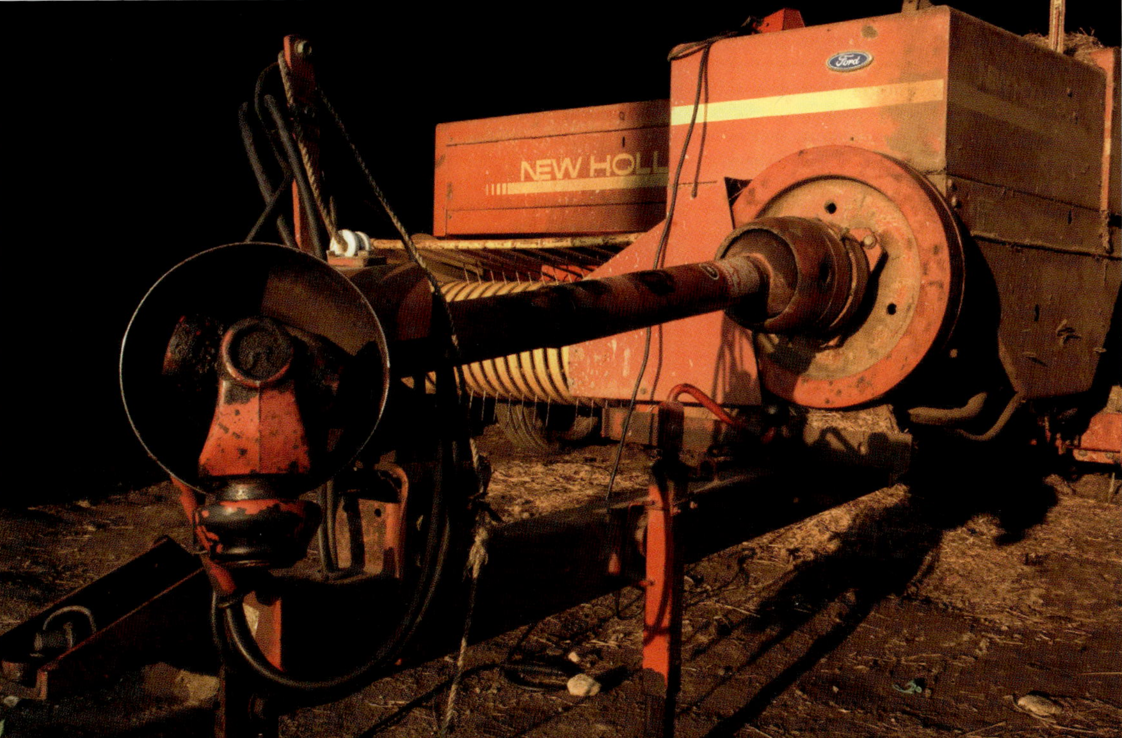

Heavy equipment in one of the newer shed barns on Proctor's farm.

New and old barn blend together along the gravel road on Baldwin Hill.

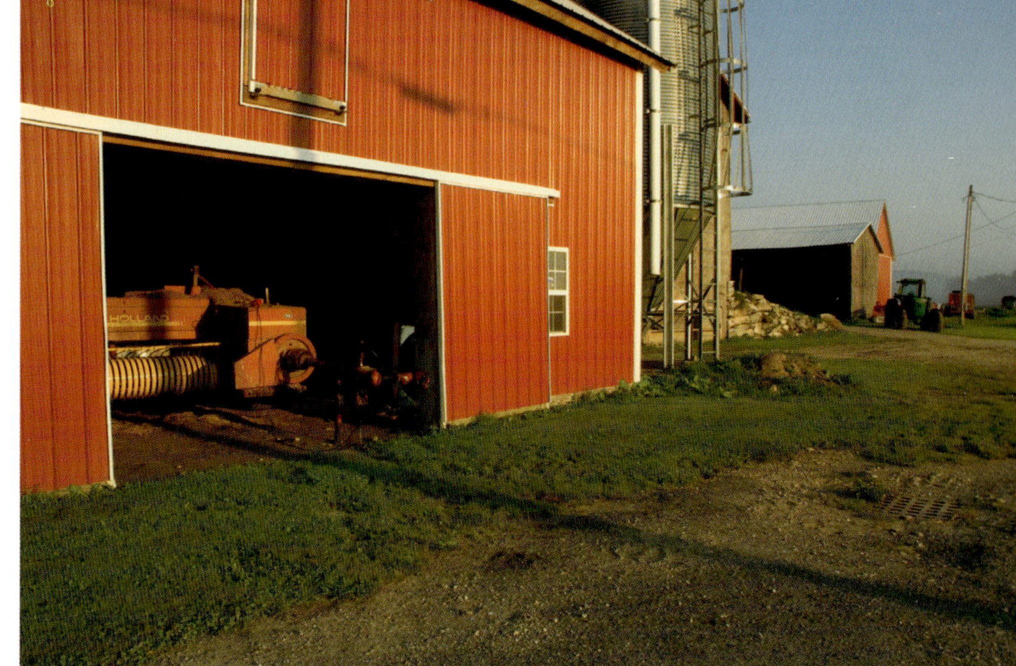

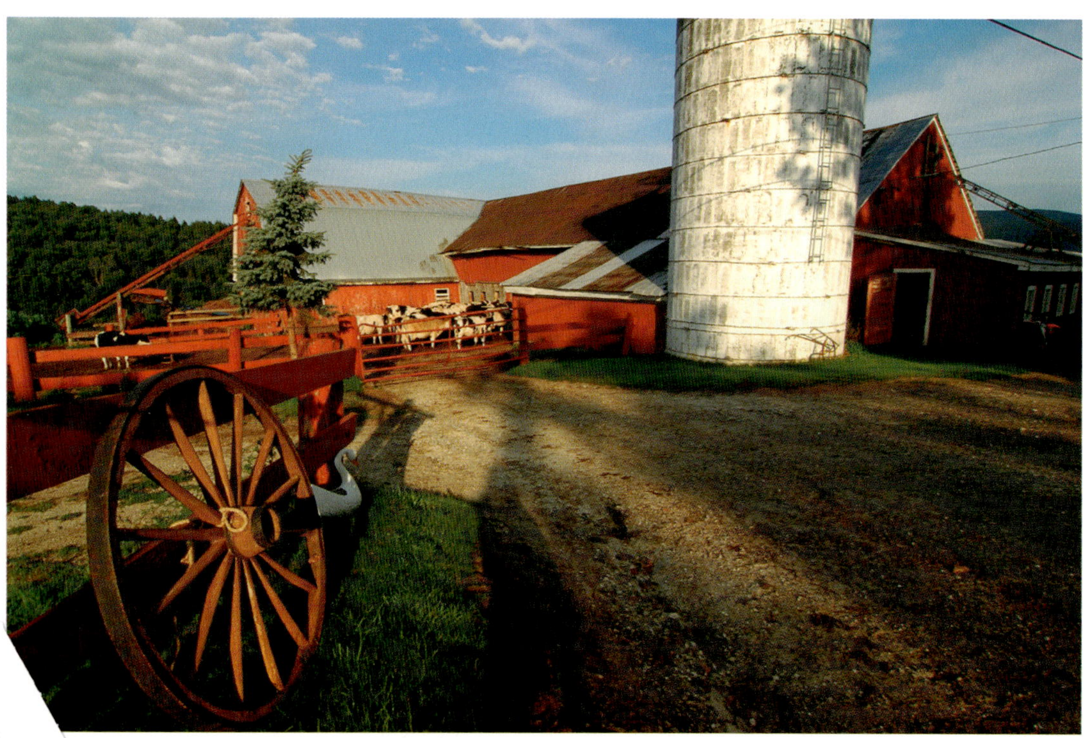

Sunrise at Alibozek's farm in Little Egypt. The gable roof side-entry barn dates to the original settlement of the farm by the Susan B. Anthony family in the early 1800s. The farm passed through many hands before the Alibozek family purchased it in 1963. The 20th century gambrel-roofed southern addition is used as a calf barn below and hay mow in the upper loft.

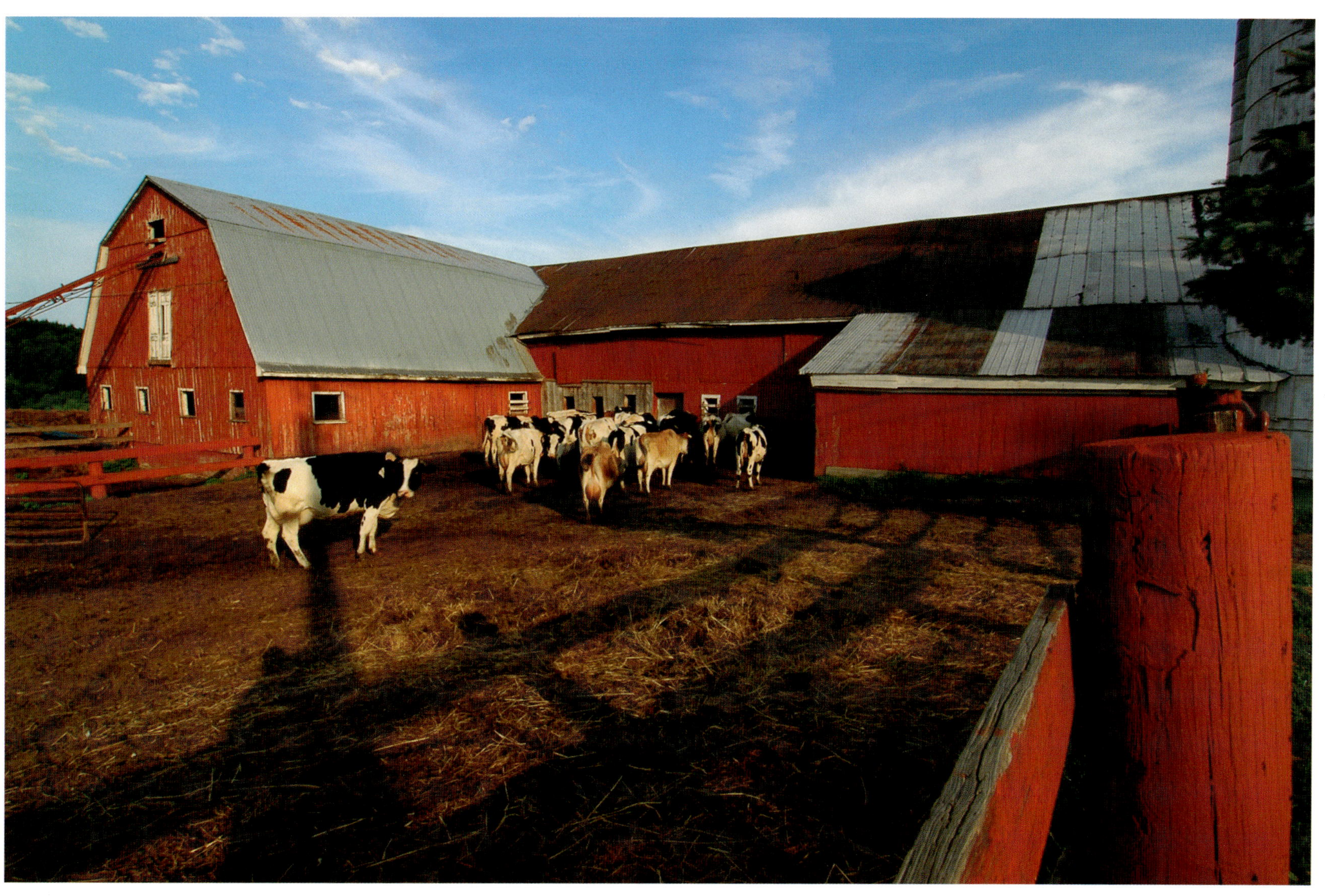

Cows being brought into the old barn for morning milking at the Alibozek farm.

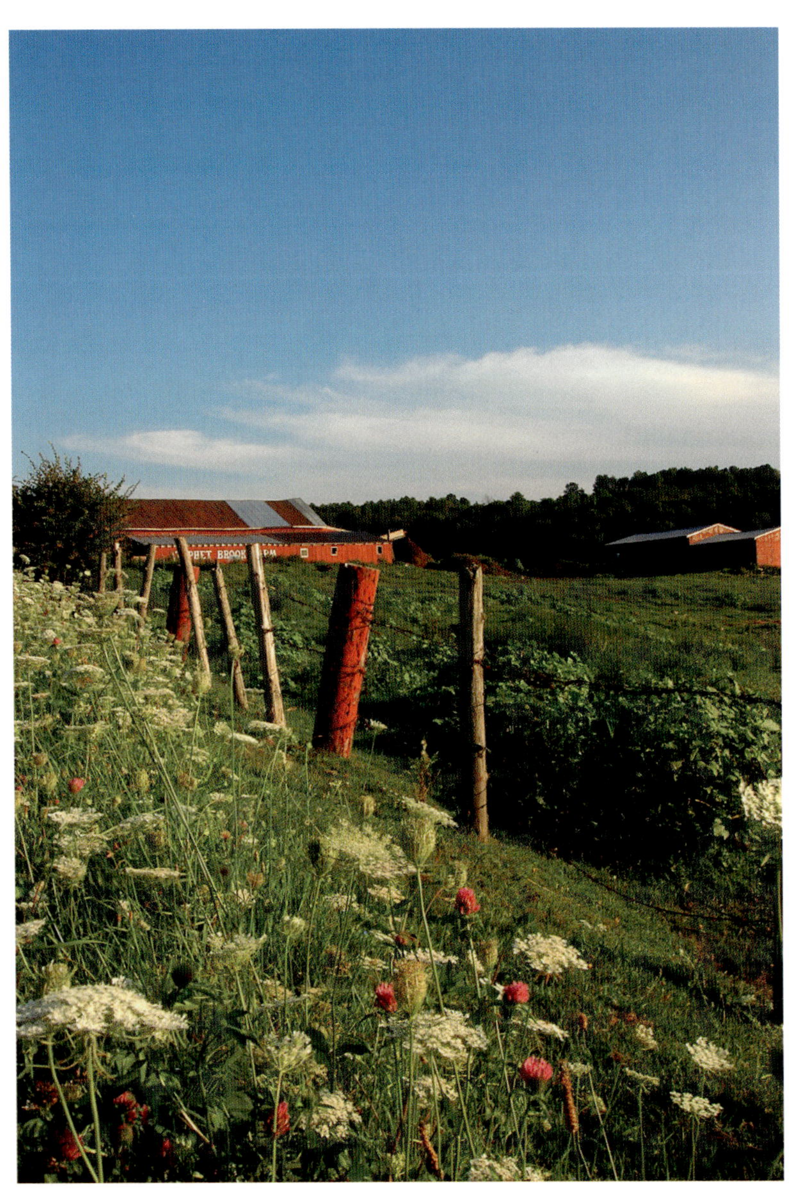

The Alibozek barns viewed from the fence line along the northern approach to the farm on East Mountain Road.

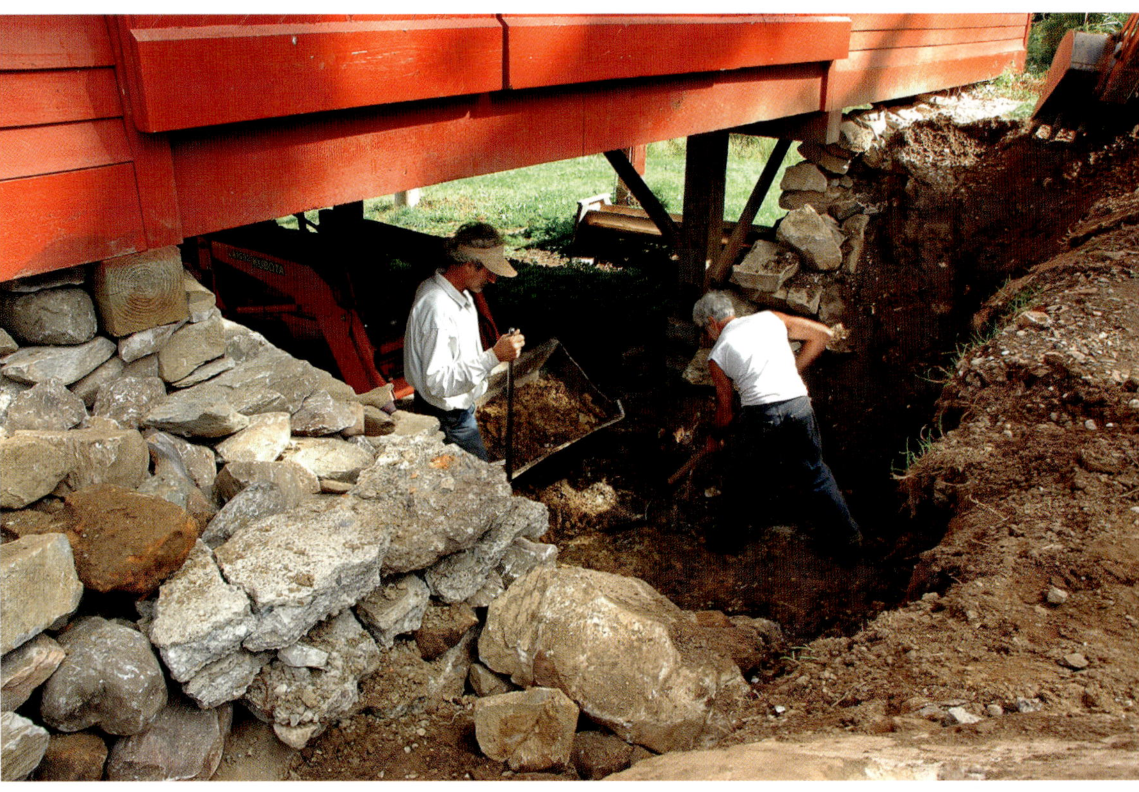

Stone foundation wall restorations underneath the great stilted red barn at Lila's Sheep Farm in Great Barrington. Jim Baker, owner of Natures Antiques, a specialty restoration business, and his brother Larry, spent the entire summer of 2007 excavating and rebuilding the full length of the 100-plus foot wall in 15-20 foot sections.

The stilted red barn, now almost fully restored, once sank as much as four feet from corner to corner because of weakened pilings and a sagging foundation wall.

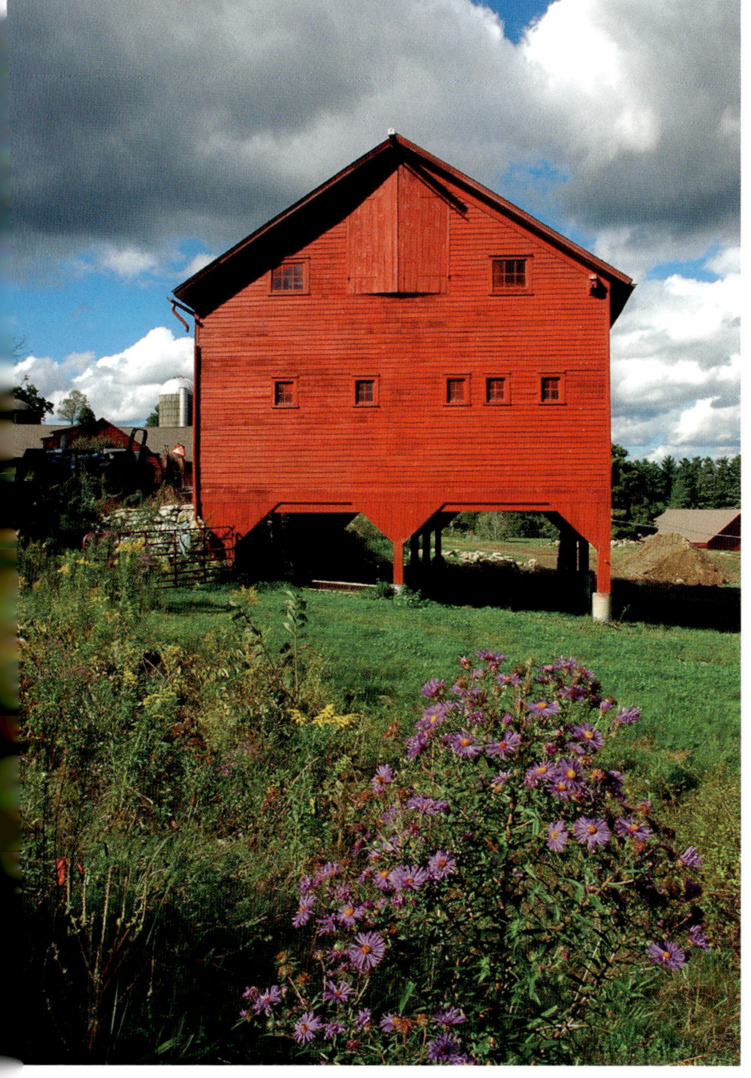

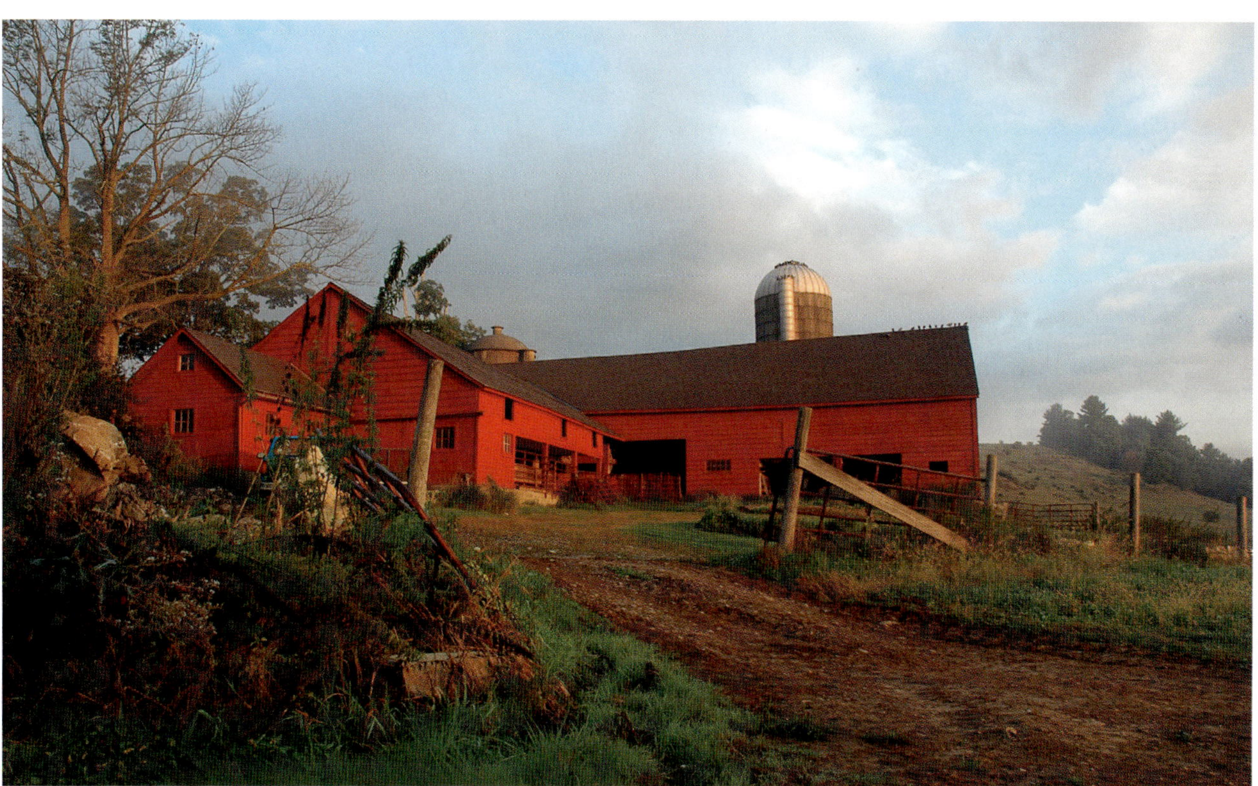

The "back barn" at Lila's Sheep Farm is the original barn and dates to around 1790. Several additions coming off of the left side of the original structure were added through the course of the 19th century. There is a square foundation inside the old barn that suggests that there was once an internal wooden silo there. According to Lila Berle, the sheep often prefer to use this barn when giving birth.

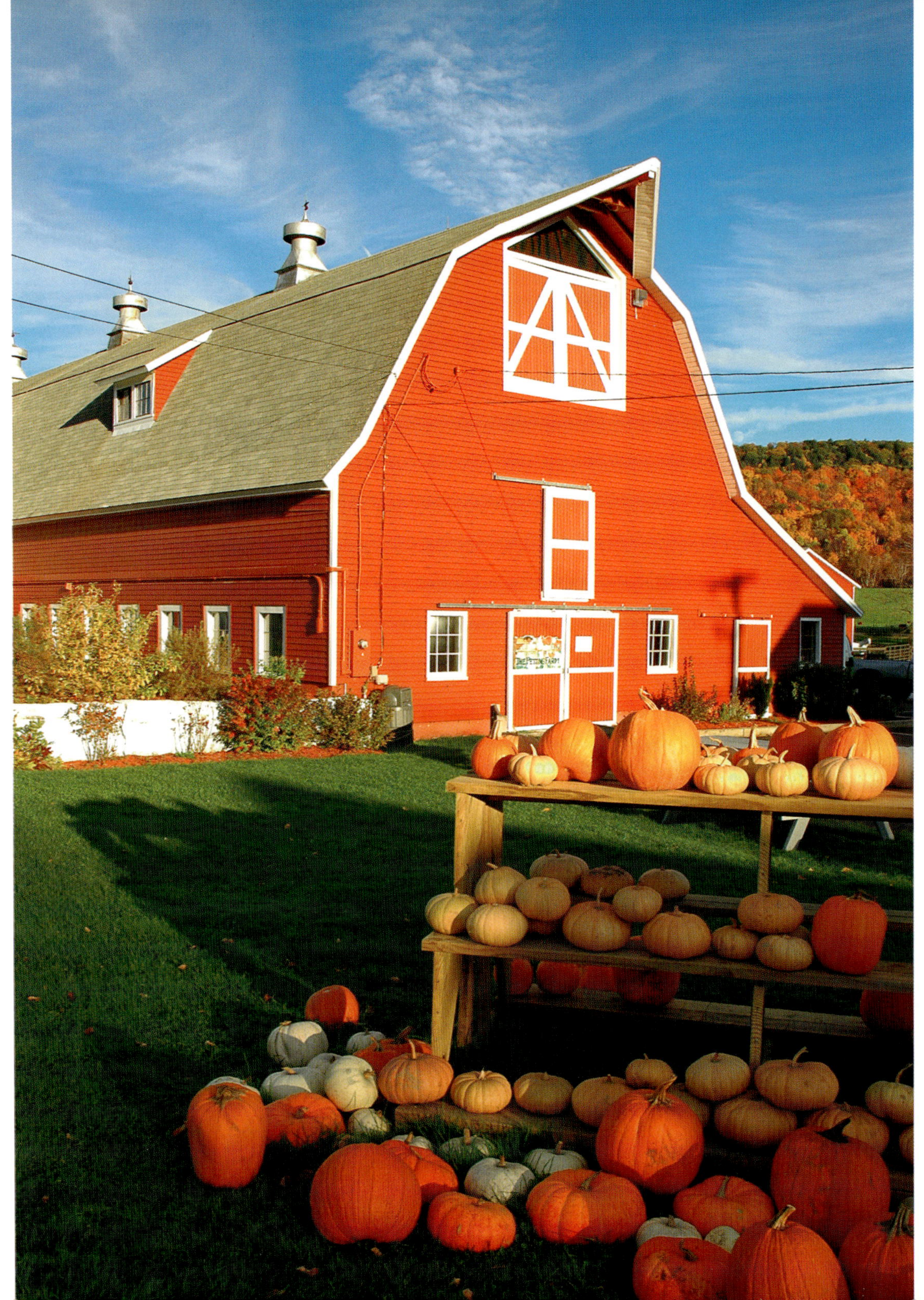

The 20th century gambrel roof barn at Green River Farms in Williamstown is known to everyone who has driven along the northern portions of Route 7. On any given day you will find numerous cars pulled over at the crest of the hill just north of Five Corners as their owners take photos of Mount Greylock and the Green River Farms below. The immaculate barn has classic red paint with white trim and is the centerpiece of the farms retail operations.

The gable front of the barn at Green River Farms showing the projecting hood on the peak over the hay mow door.

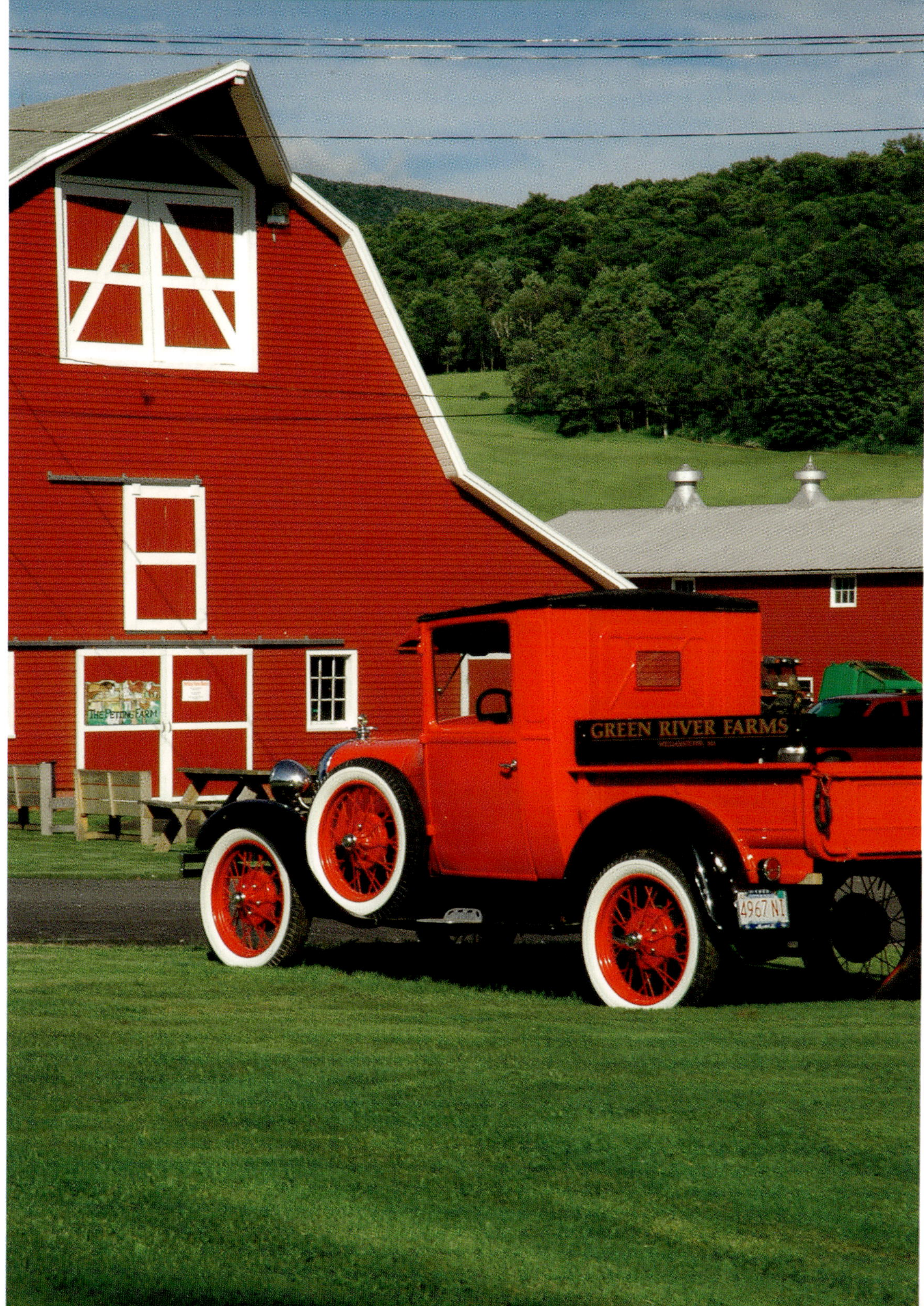

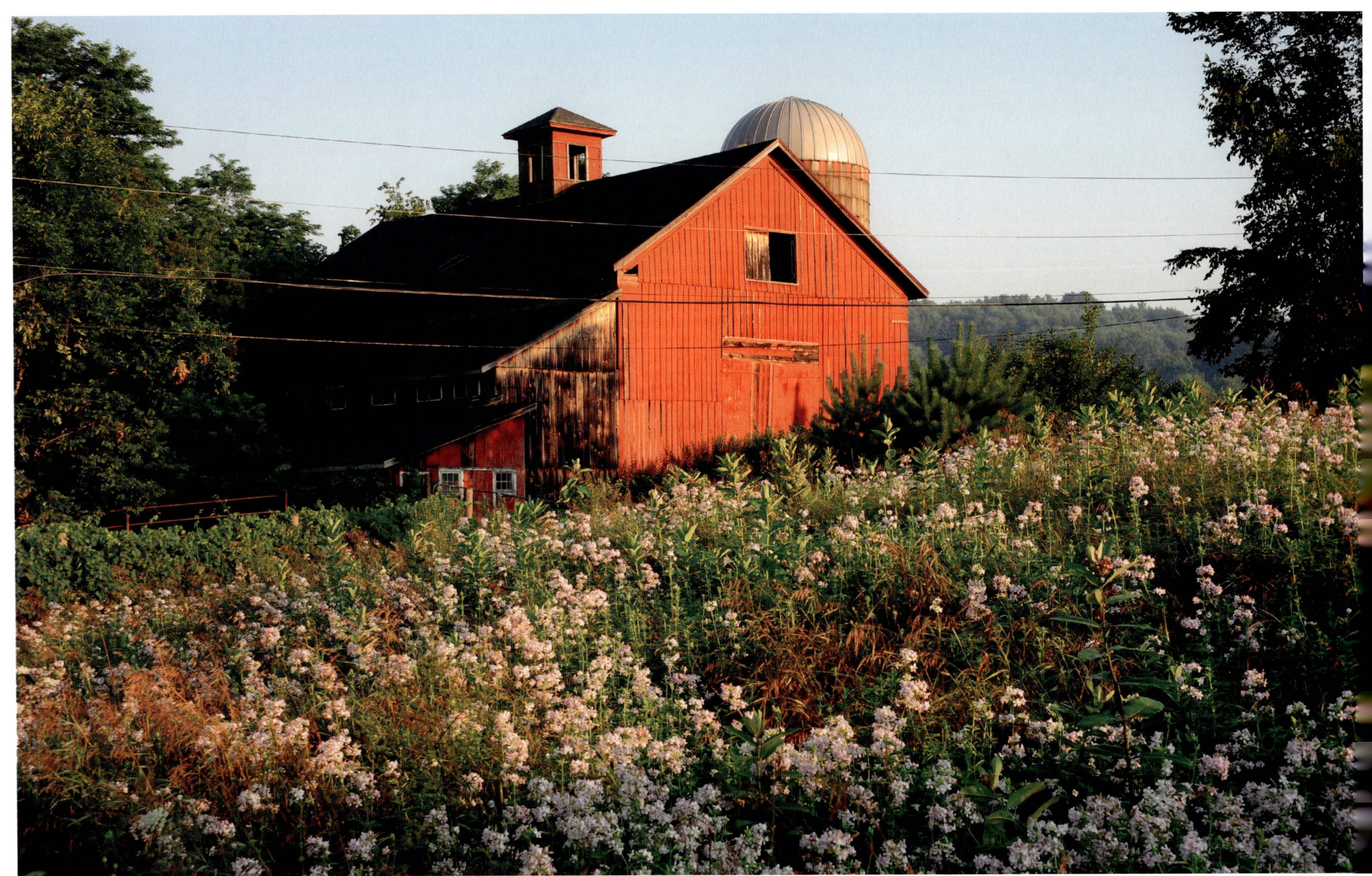

This nineteenth century gable-entry barn in Mill River has few signs of activity.

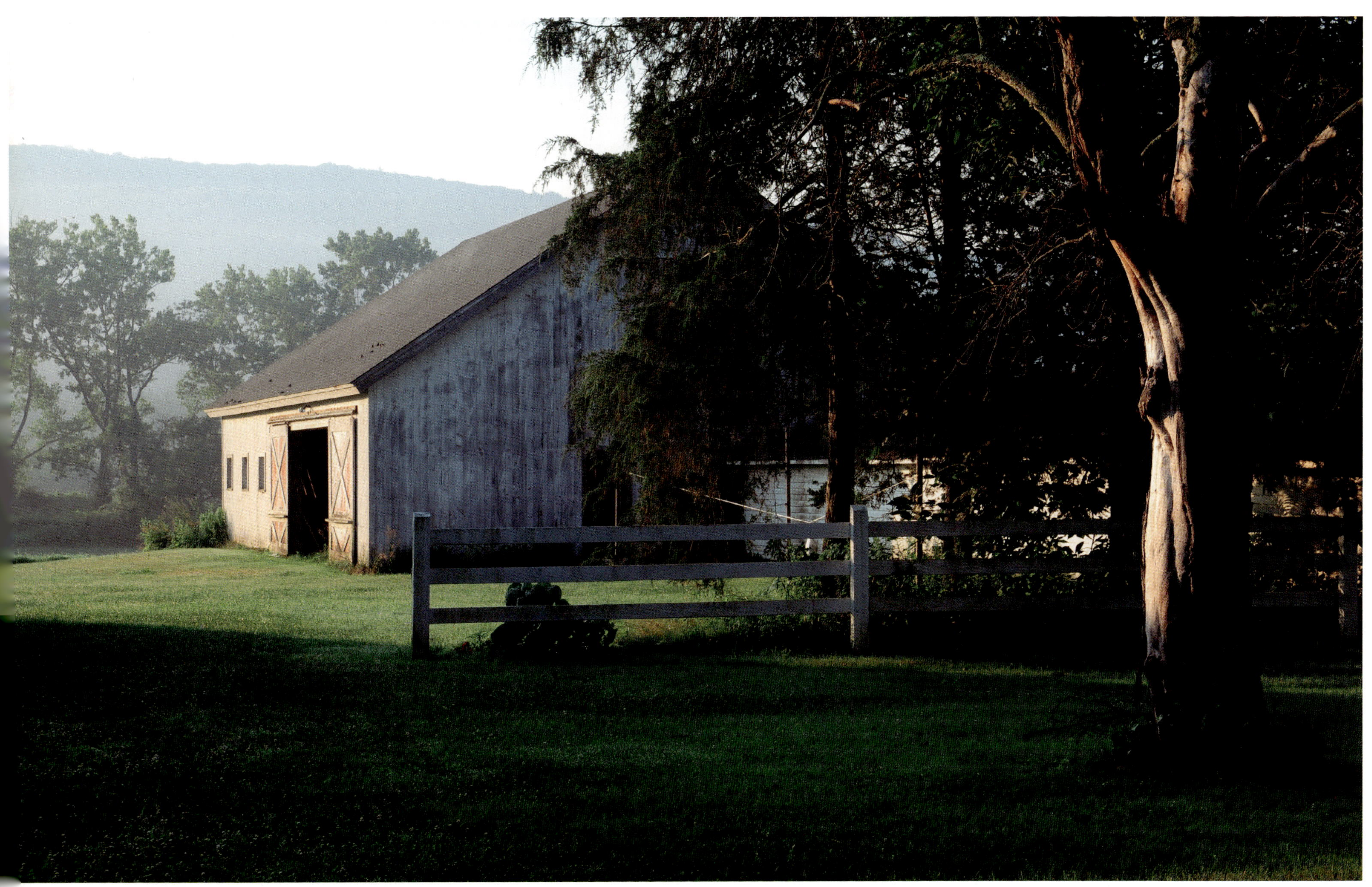

In another instance of a drive-by shooting, I spotted this barn at sunrise while driving to a different location in Clayton and spent only a few minutes getting this image in the first light of day.

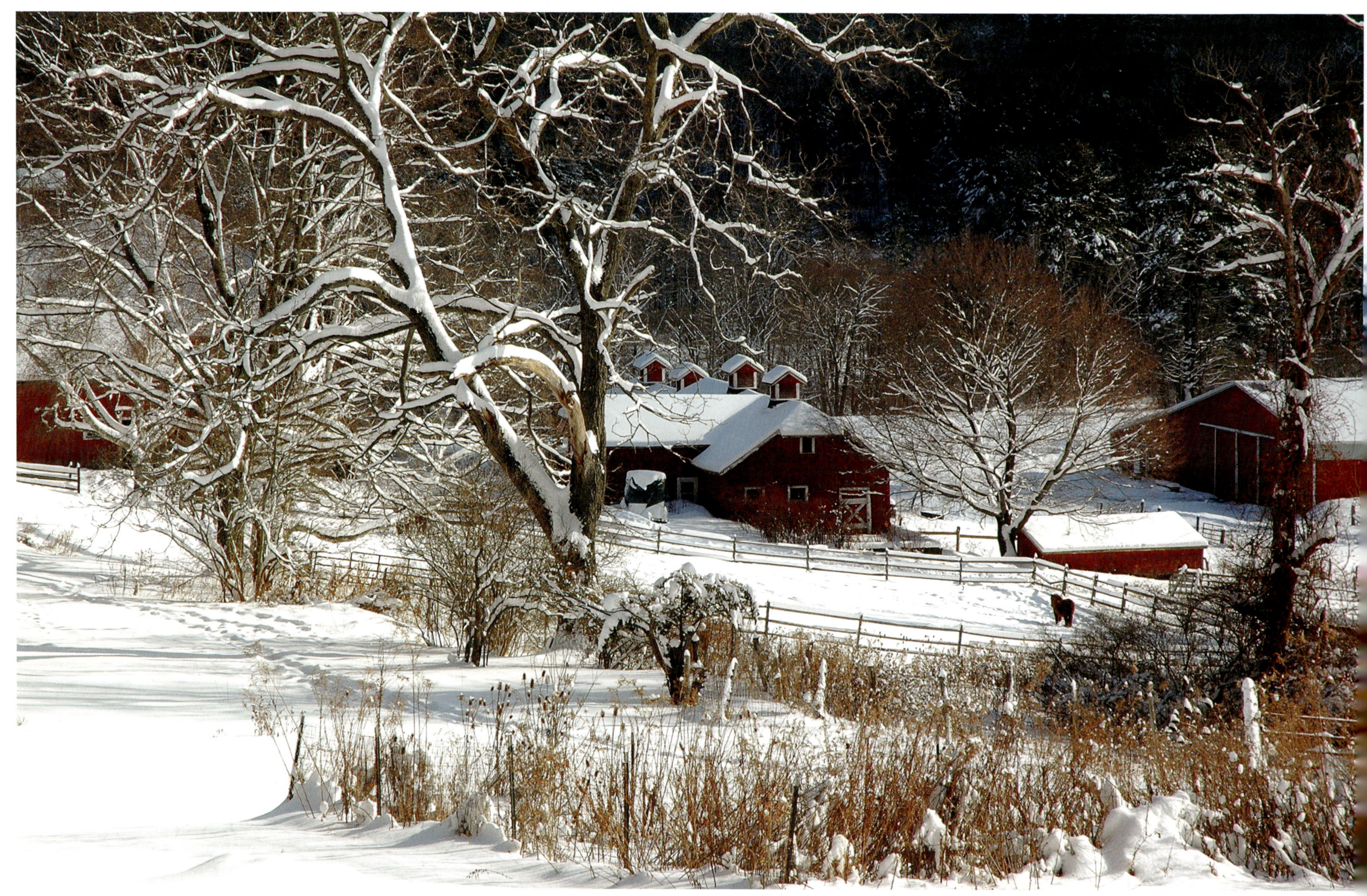

Taken form Under Mountain Road in Lenox these barns, part of Undermountain Farm, had a postcard look in the early morning after a snowfall.

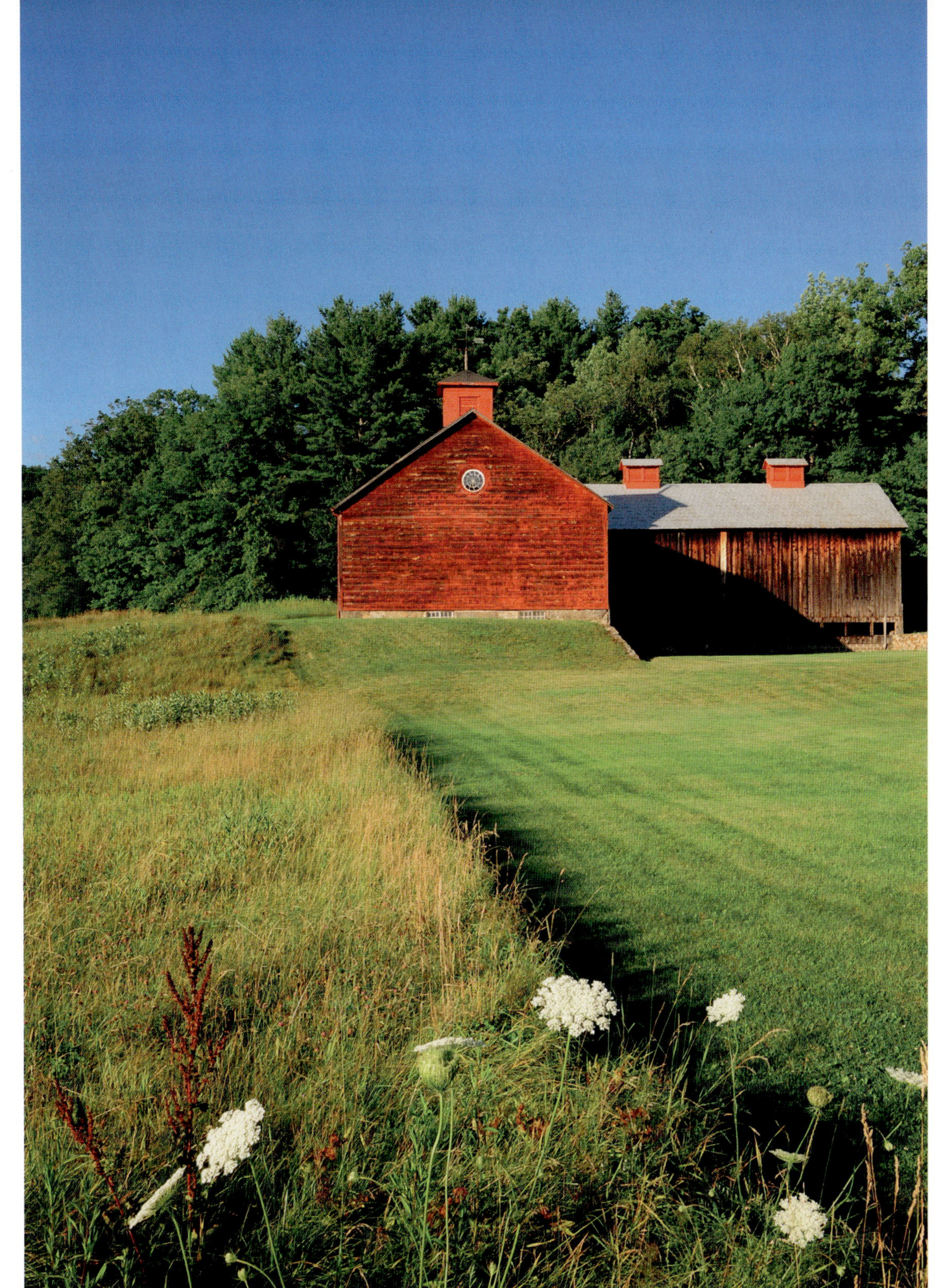

Another vantage point of the iconic spiral-windowed barn on Main Street in Tyringham.

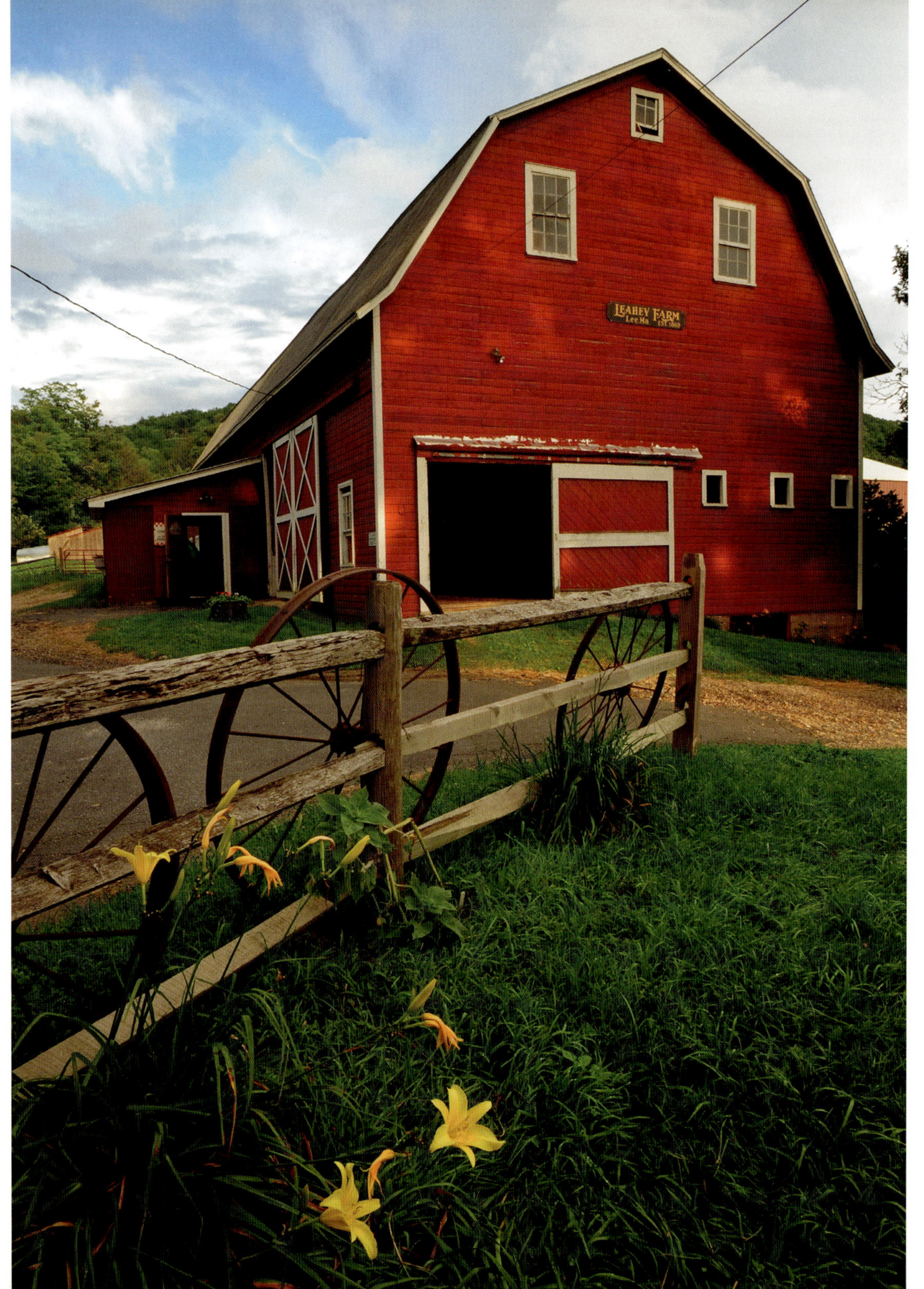

The front, west-facing end of the 1934 barn at Leahey Farm in Lee with a sliding door leading into the horse stalls in the lower front area of the barn.

Right:
Old milk bottle crates stacked up in the loft of the Leahey barn.

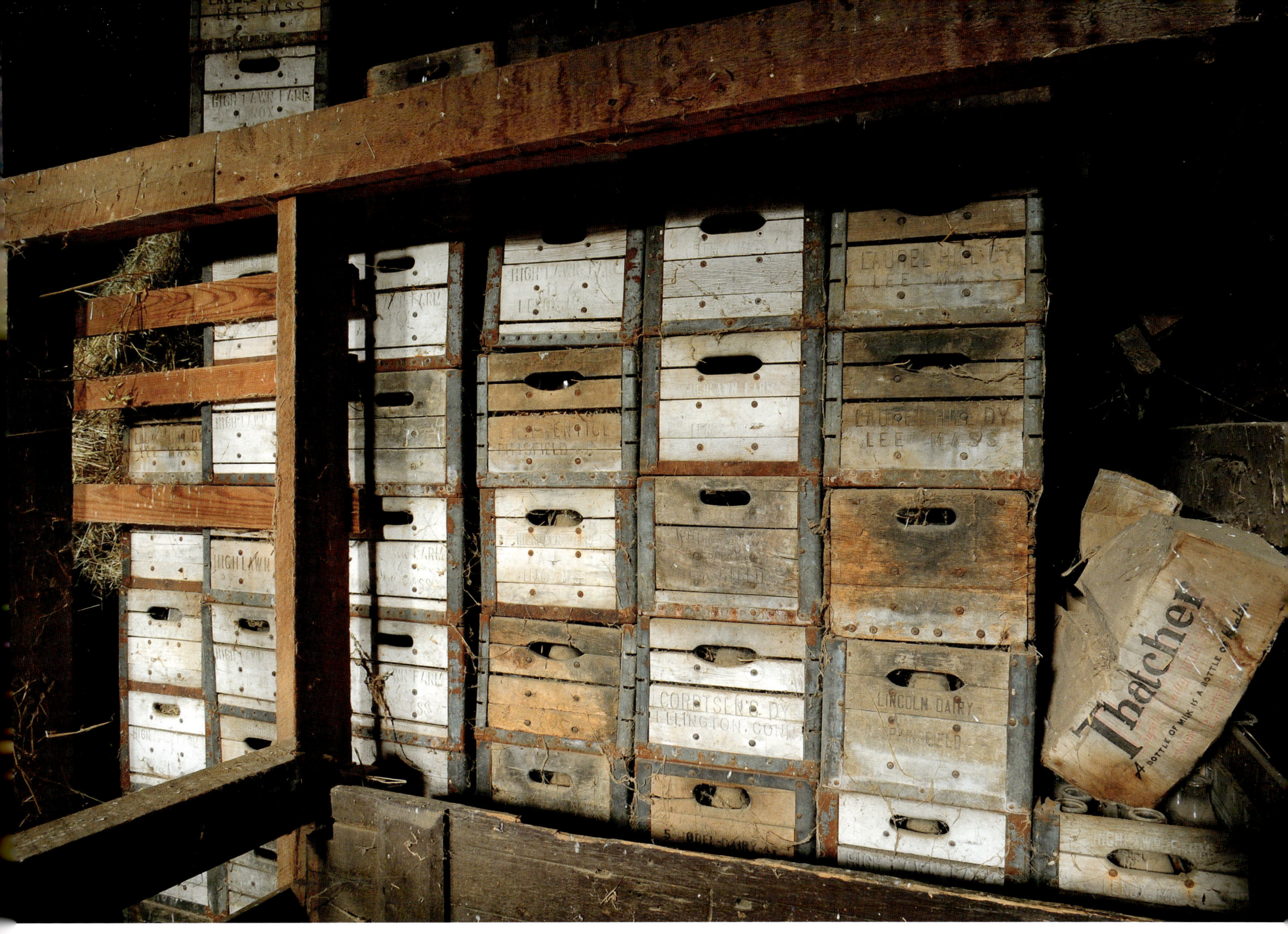
LAUREL HILL DY
LEE MASS
HIGH LAWN FARM
LINCOLN DAIRY
Thatcher
A BOTTLE OF MILK IS A BOTTLE OF

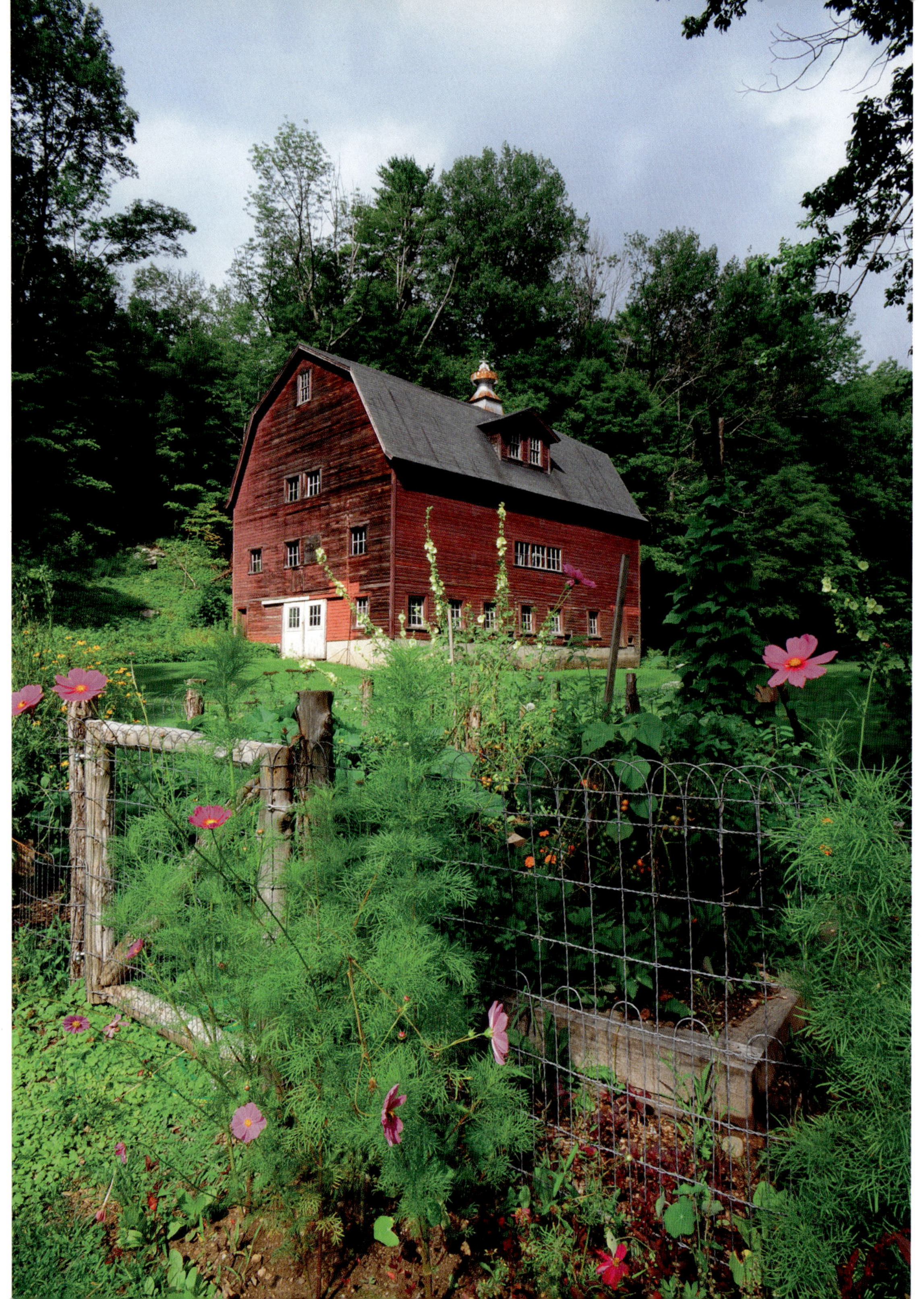

Early 20th century barn on Lake Buel Road near Hartsville in New Marlborough. While not a side-hill or bank barn by definition, this barn is connected by a bridge in the back to a meadow above this part of the property.

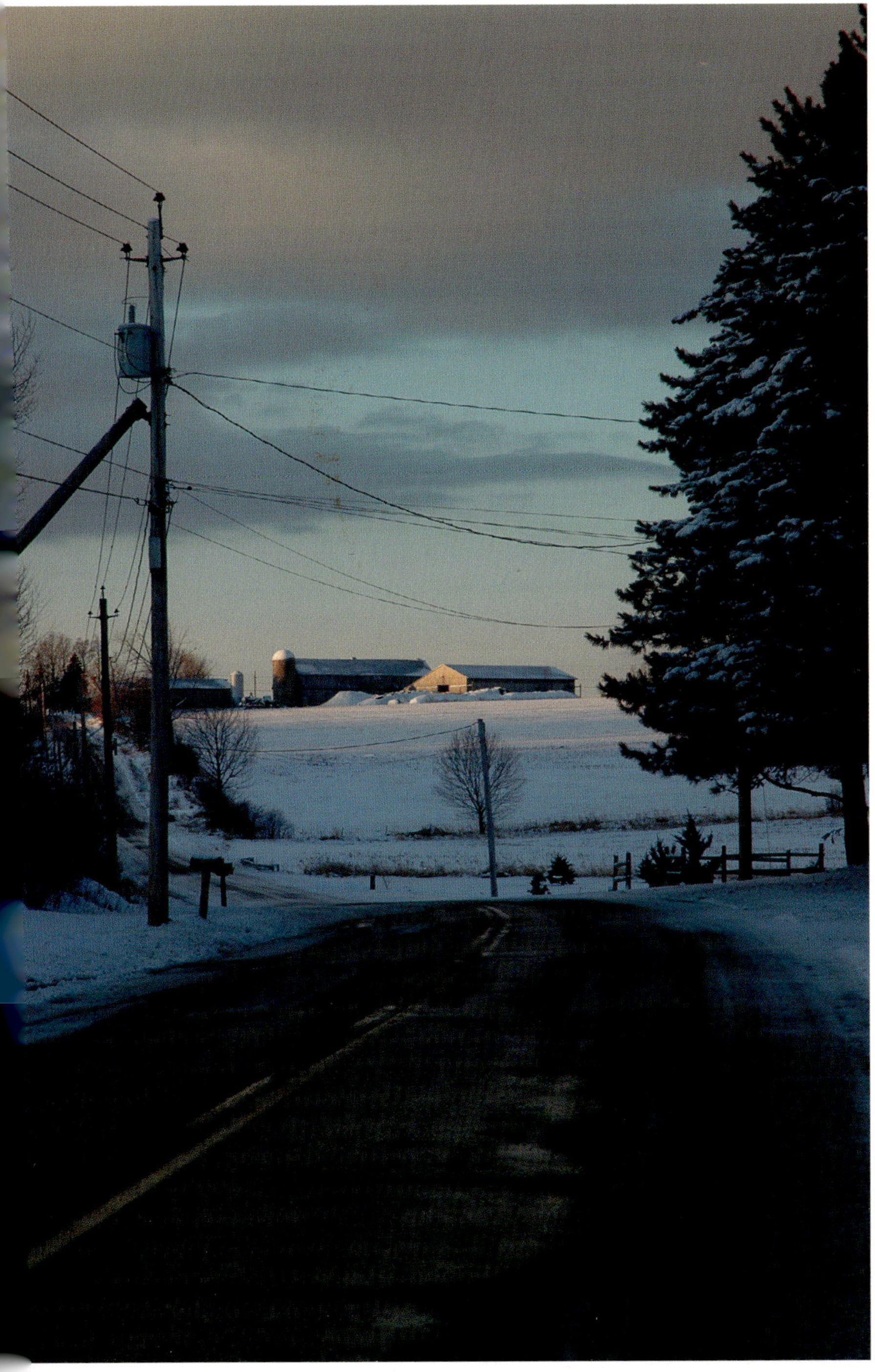

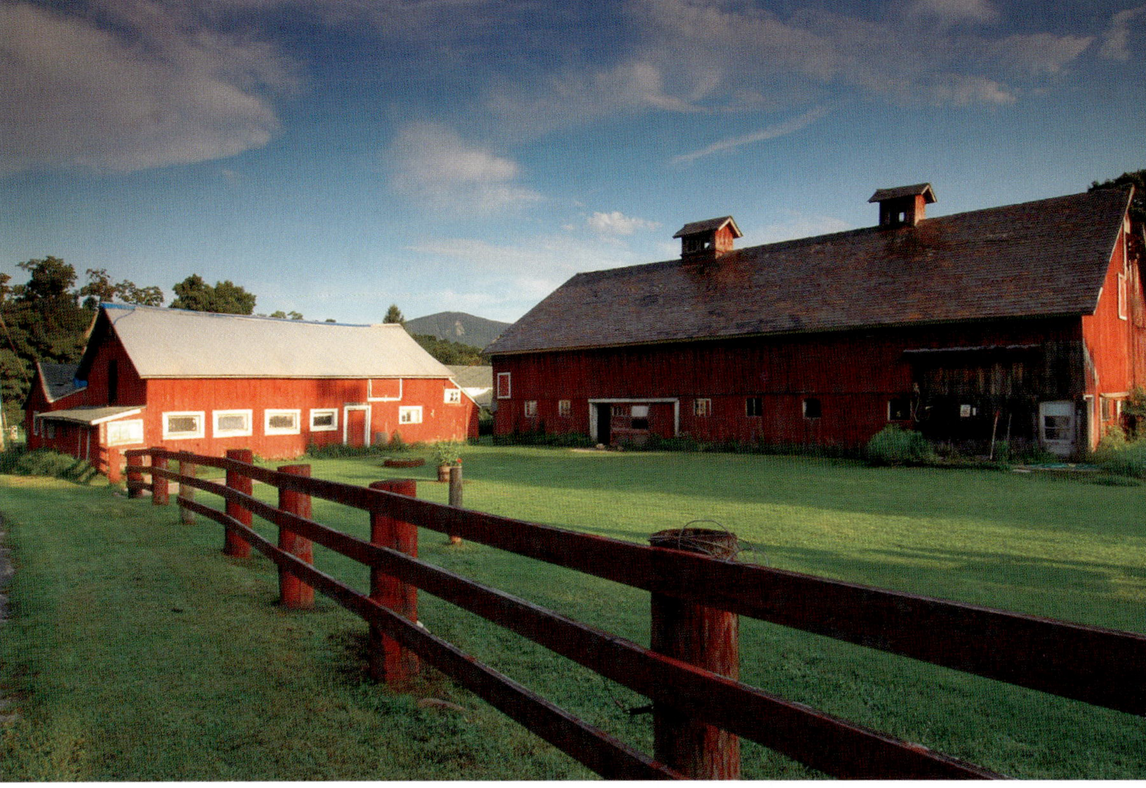

These two barns, featured on the book's cover, stand on Burlingame Hill in Adams. The hill is named after the family that originally settled the land, and the farm is now called Golden Acres Farm for the rich light that adorns it late in the day. The smaller English barn dates to 1782, the larger 40 ft. by 100 ft. barn caused a sensation in the area when it was built. It was famously dedicated on July 4th, 1830 by Elder Leland, the famous pastor and purveyor of Cheshire Cheese, in front of a large gathering of townspeople from Adams and the surrounding towns.

Sunrise along Wells Road in Cheshire on a frigid winter morning provided this image of Ted Jayko's barns. These 20th century barns were built with beams, some over 50 feet long, that were recovered from decommissioned mills in Adams.

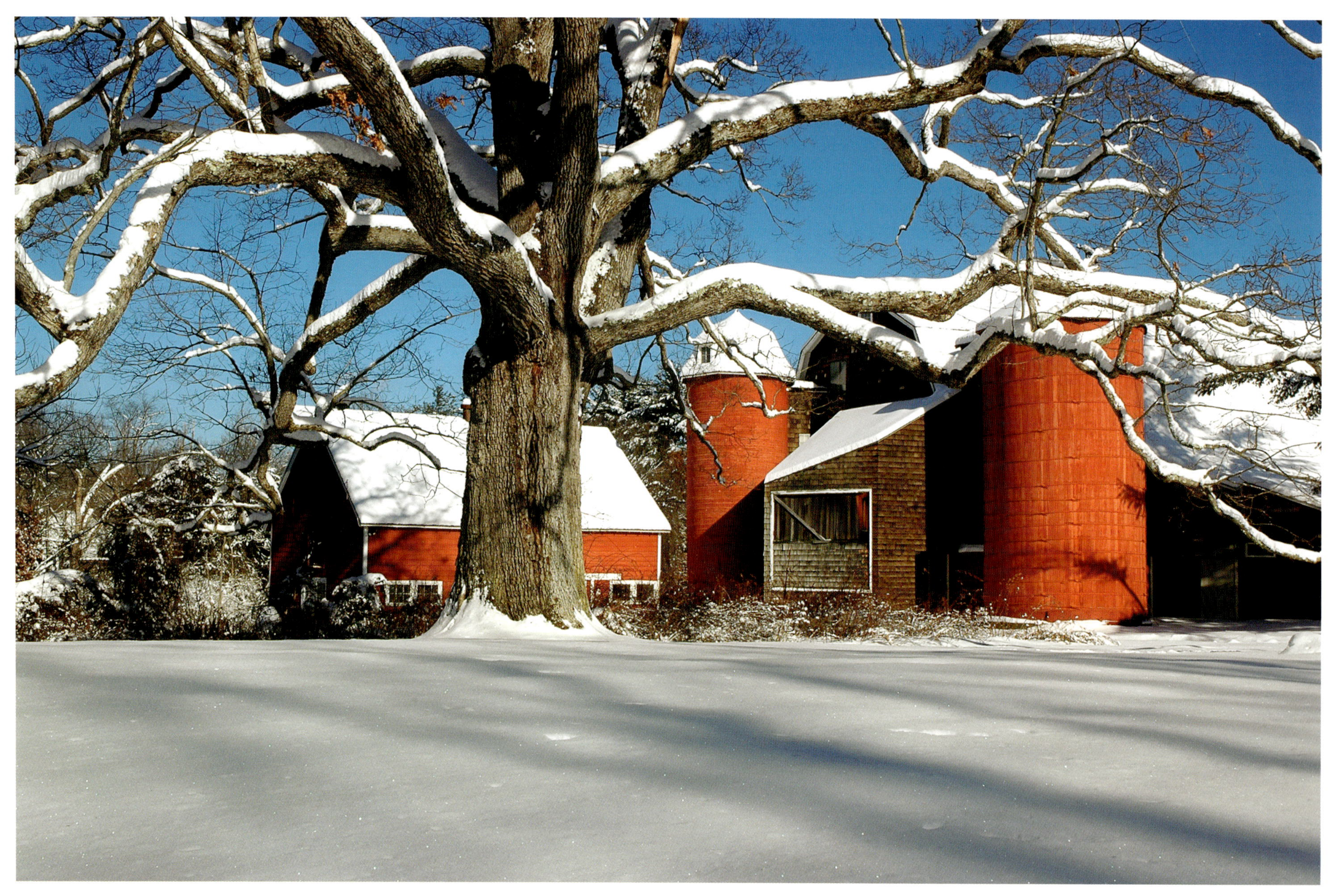

An impressive cluster of red barns and silos at Simon's Rock College of Bard in Great Barrington has been converted to an arts studio and performance space. This massive tree towers over the southwest corner of the complex.

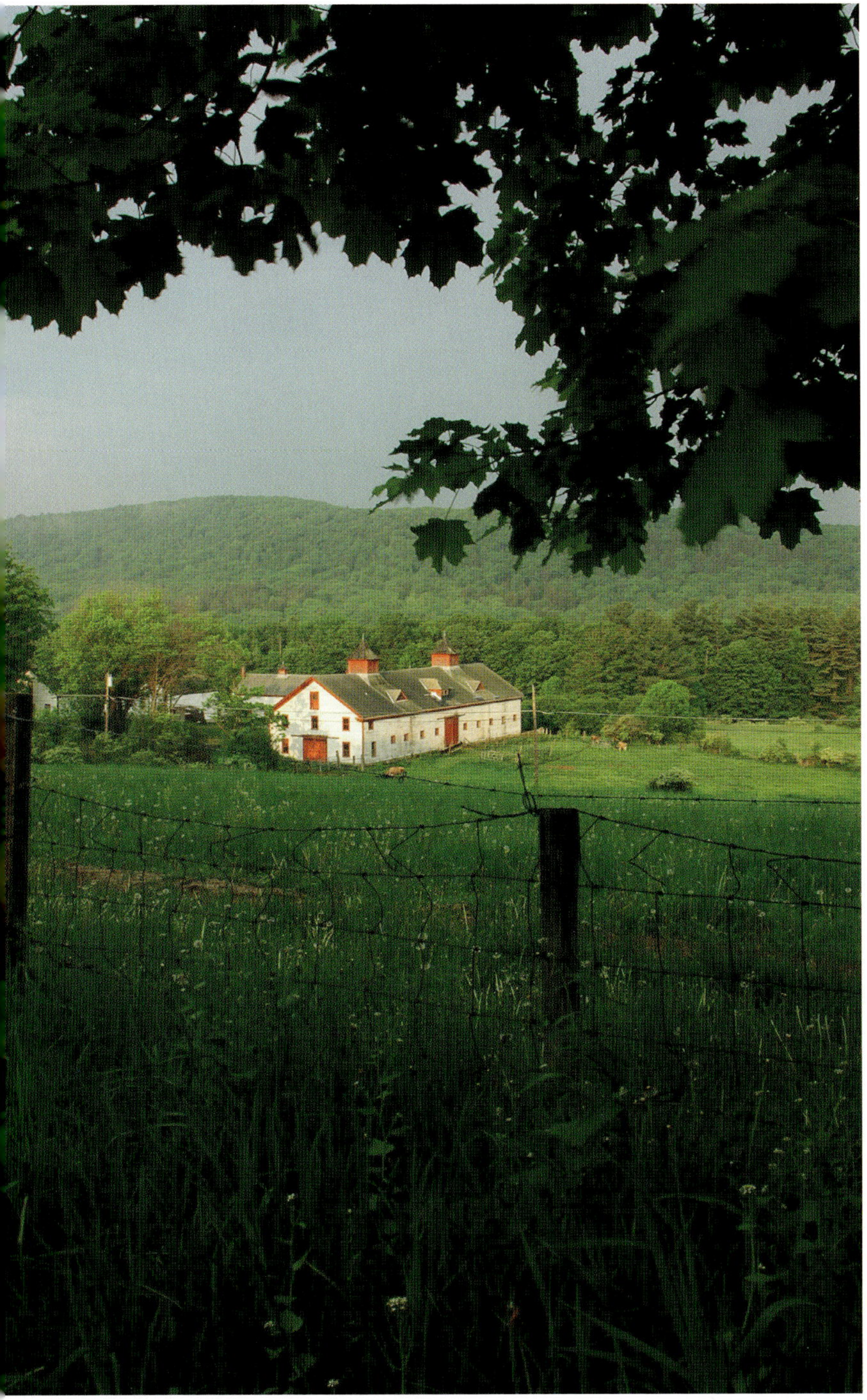

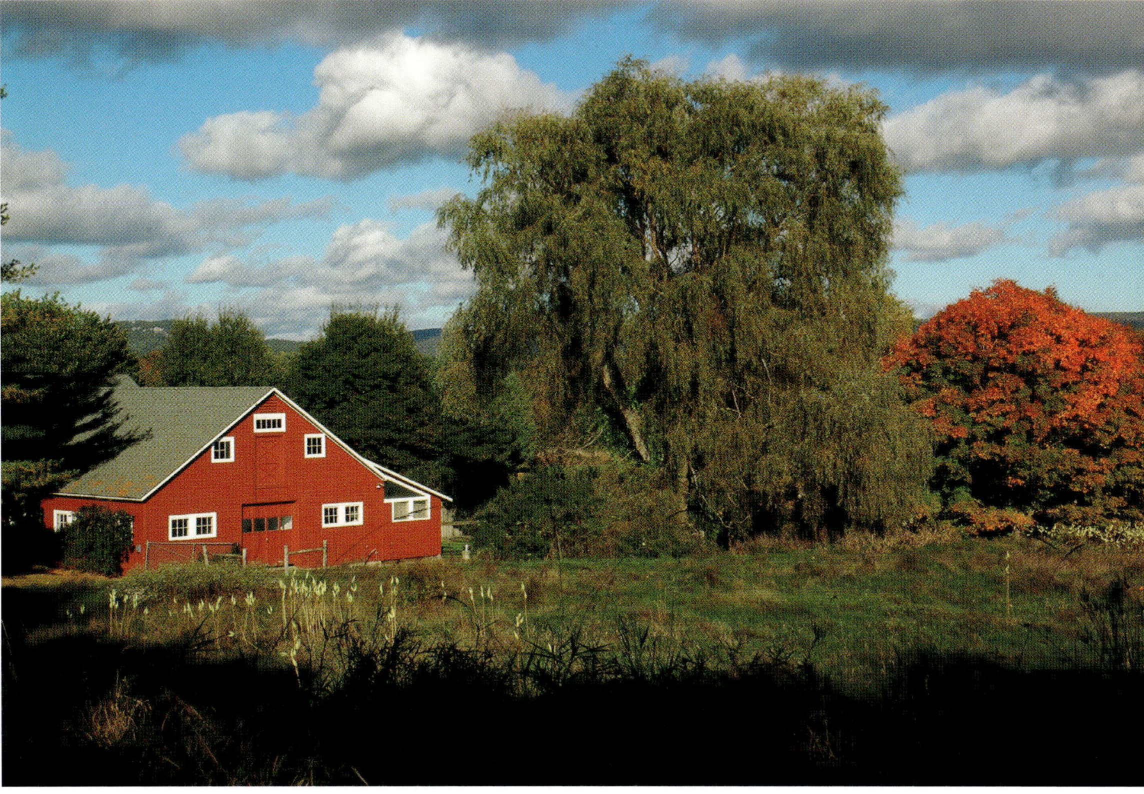

A converted barn that now serves as part of a bed and breakfast on Round Hill Road in Great Barrington has Monument Mountain, which can be seen here just above the roof line of the barn, as its backdrop.

The lower dairy barn at High Lawn Farm in Lenox at sunrise. This is another building that many people traveling along Route 7 through the Berkshires see and consider an icon of the region.

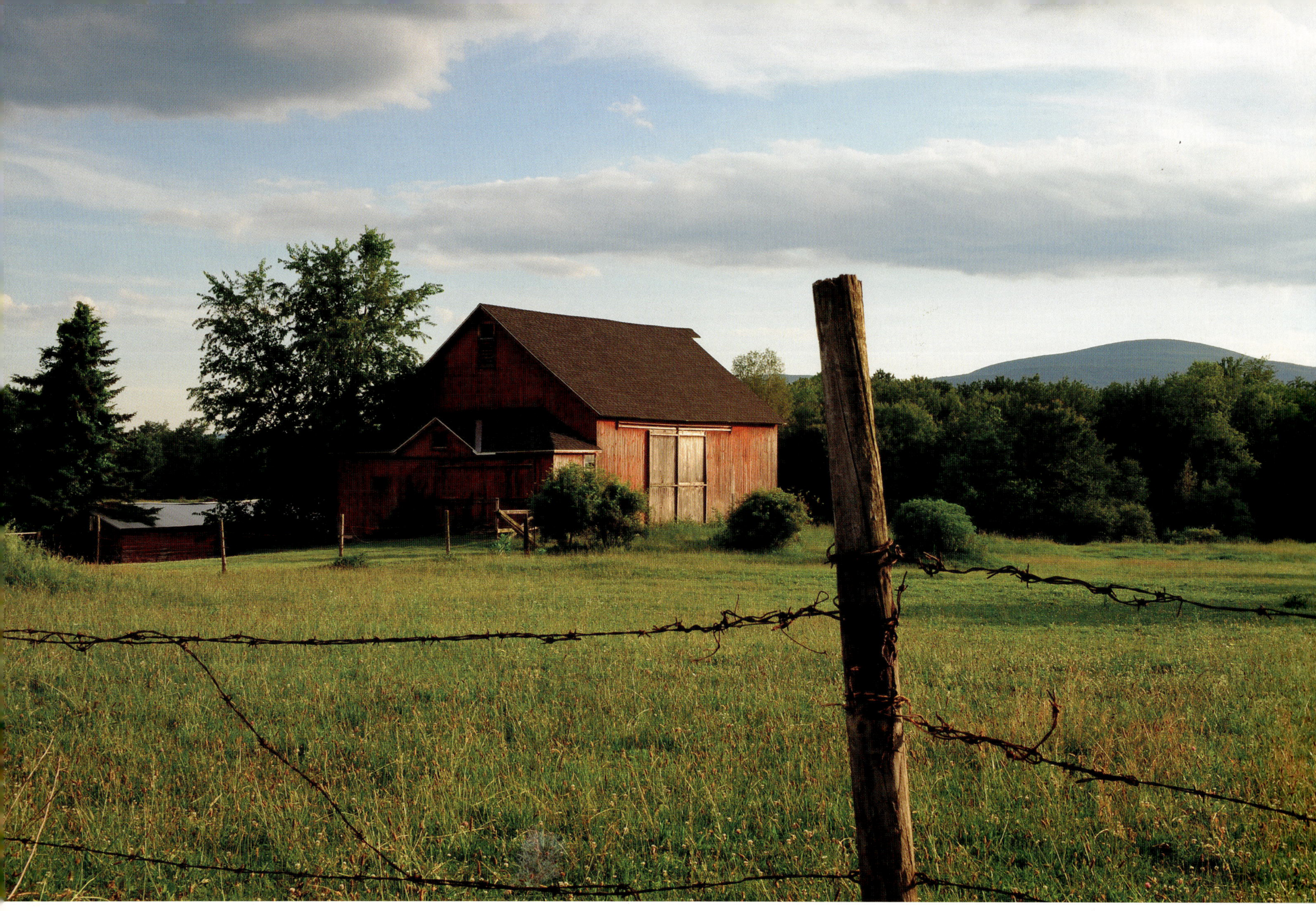

An English side-entry barn on Boardman Road in Sheffield at sunset, with Mount Everett State Reservation in the background.

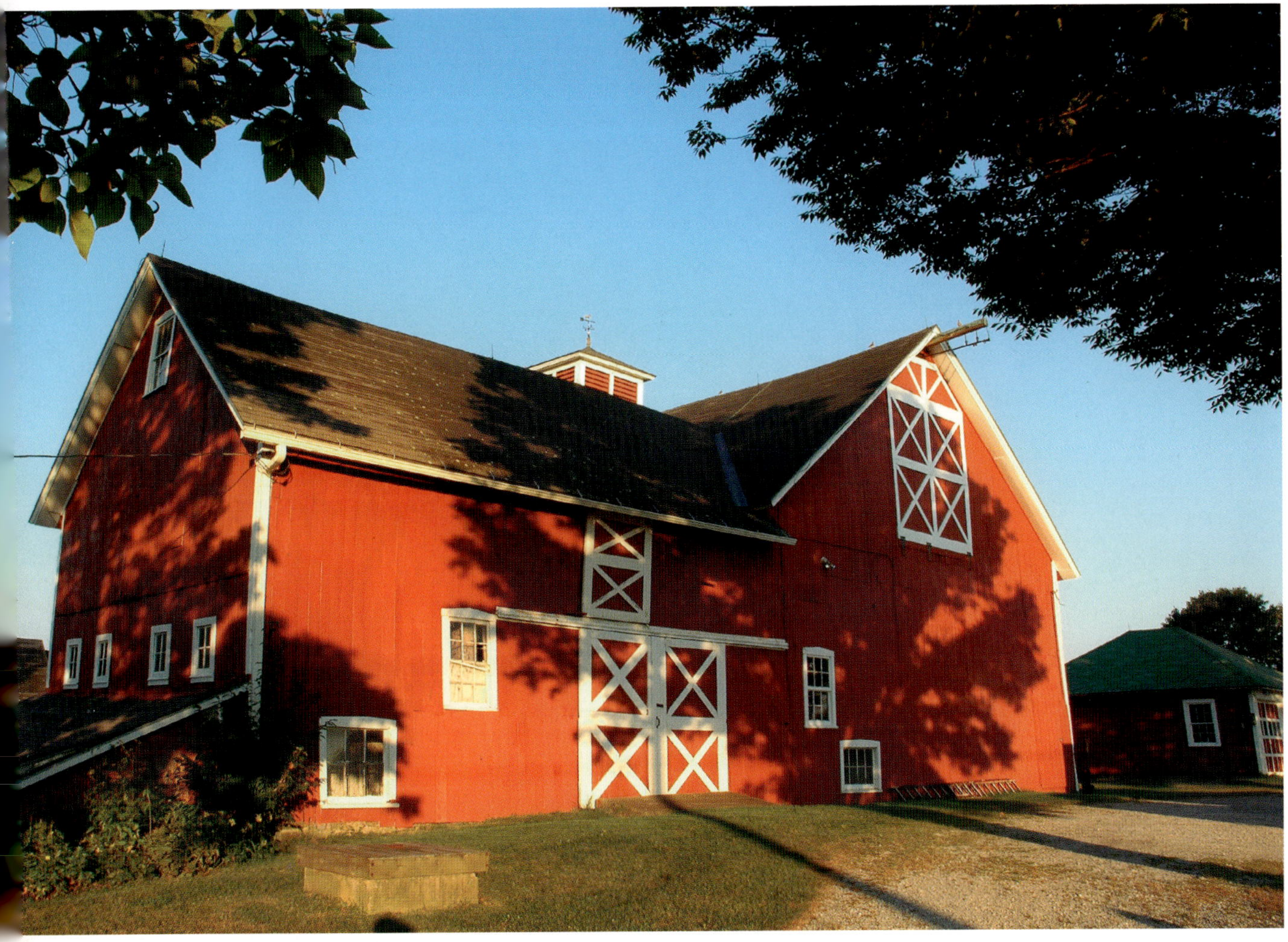

Once a dairy farm, Baldwin Hill Farm has gone through several incarnations over the last 20 years and is now a private residence. The barn has the type of red oxide painted look, with white trim accents, that became popular in the mid-to-late 1800s.

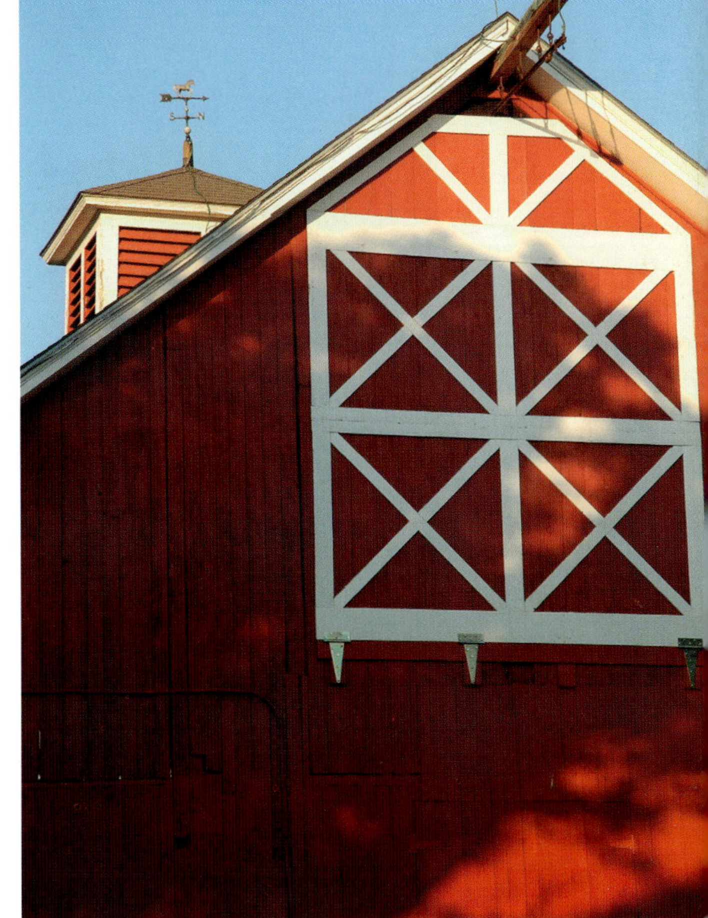

Detail of the gable-end hay loft door and the top portion of the decorated cupola on the Baldwin Hill Farm barn.

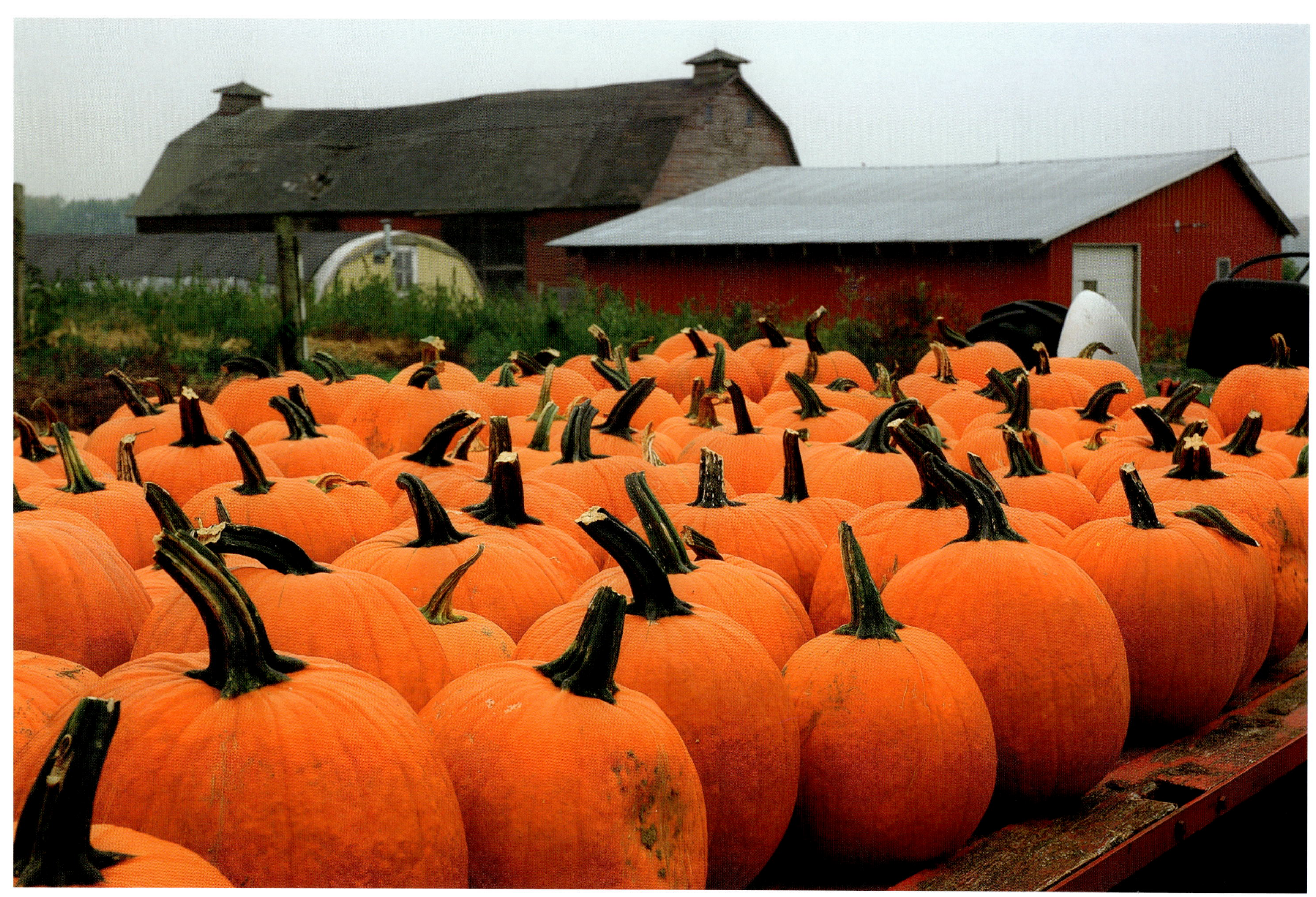

Harvest time at Boardman's Farm on Hewins Street in Sheffield.

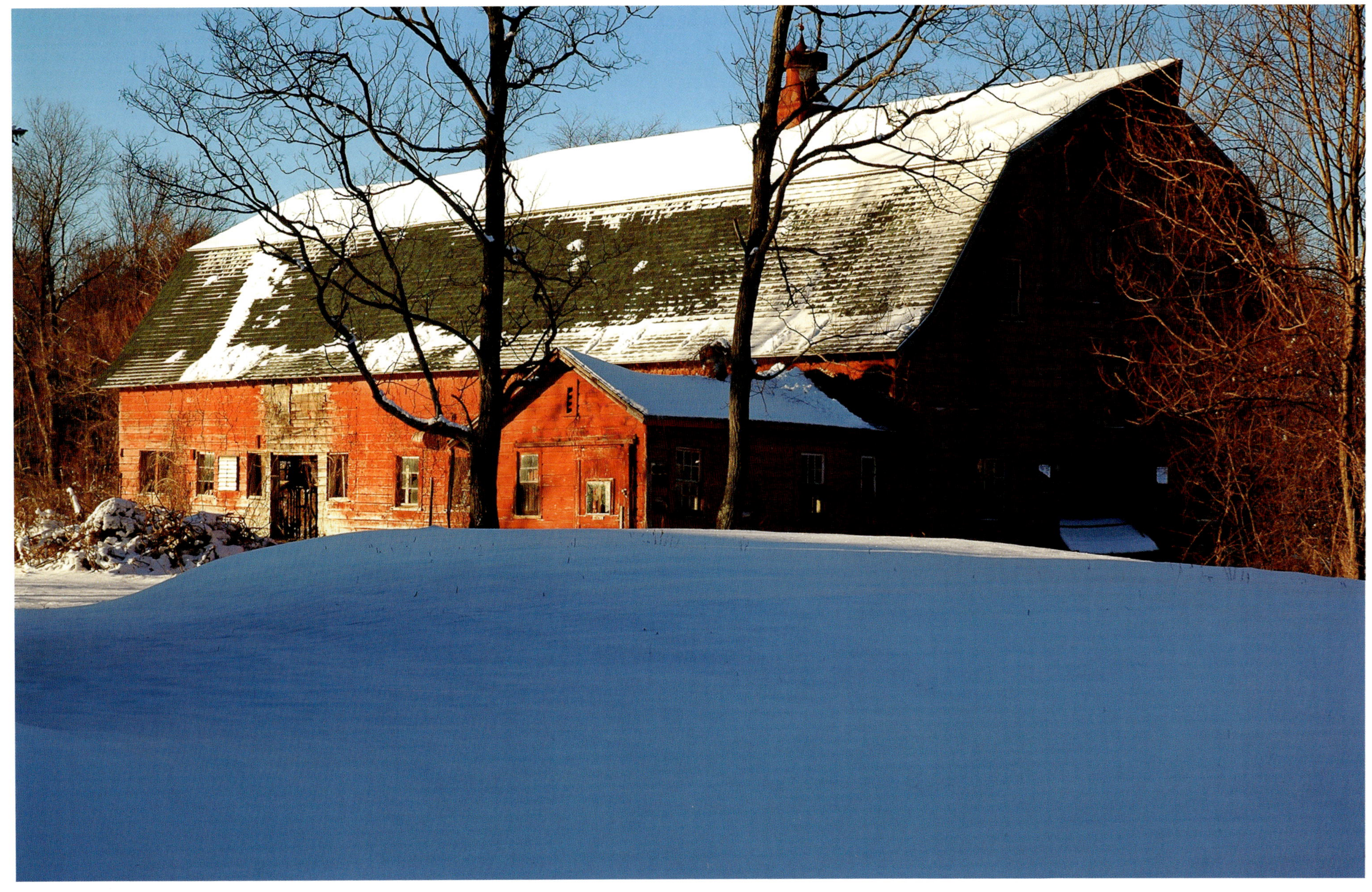

The decommissioned dairy barn on Salisbury Road, in Sheffield, seen here in the winter of 2007, is almost certainly destined for demolition. The large barn has beautiful lines including the projecting hood on the gambrel roof peak.

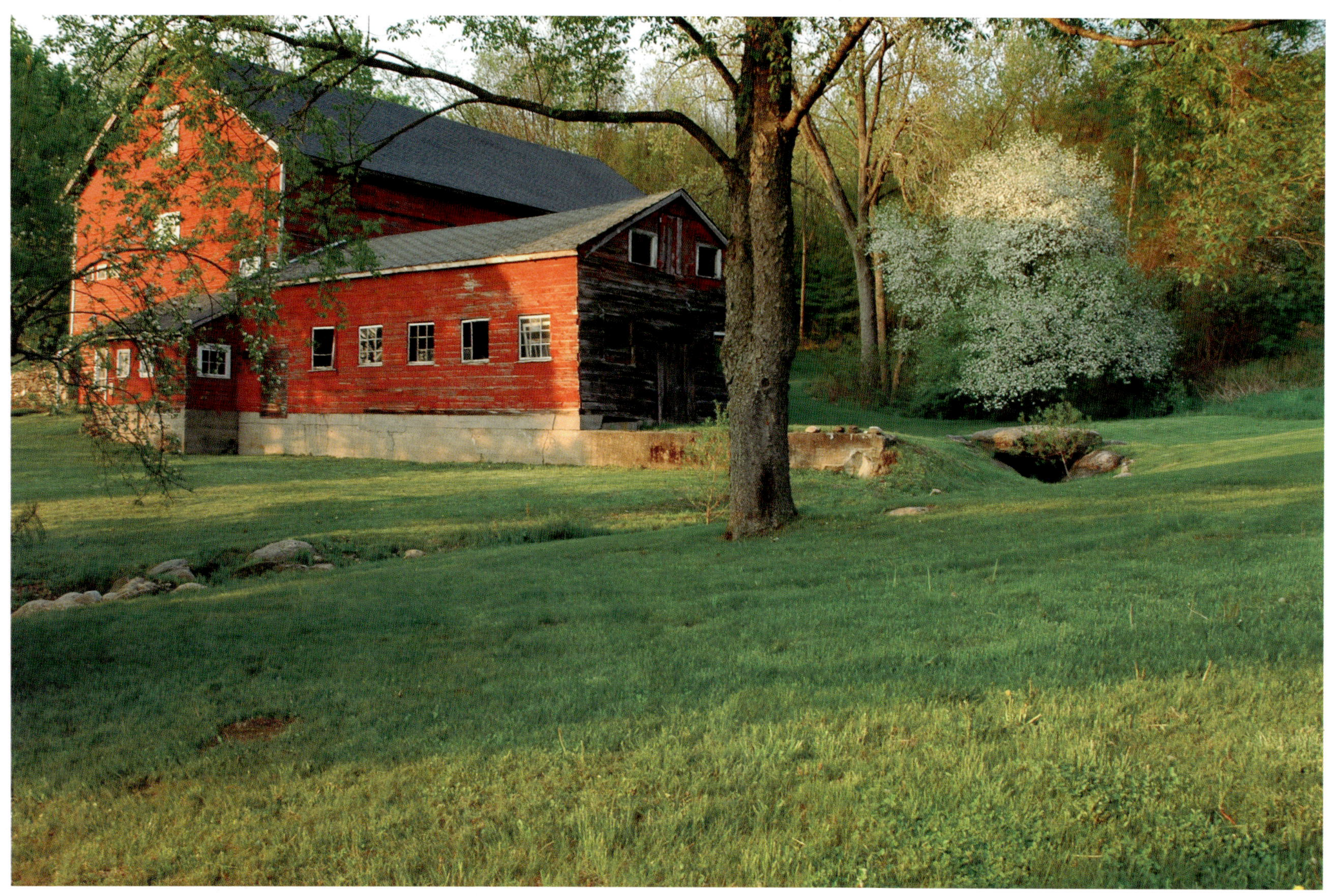

A barn converted to a private residence in Sheffield.

Right:
A barn with a shallow domed roof line in Tryingham. Most likely a 20th century barn, this has an unusual design for this region and I cannot recall seeing another barn like it in my travels for this project. This was a drive-by so I have no information about the barn.

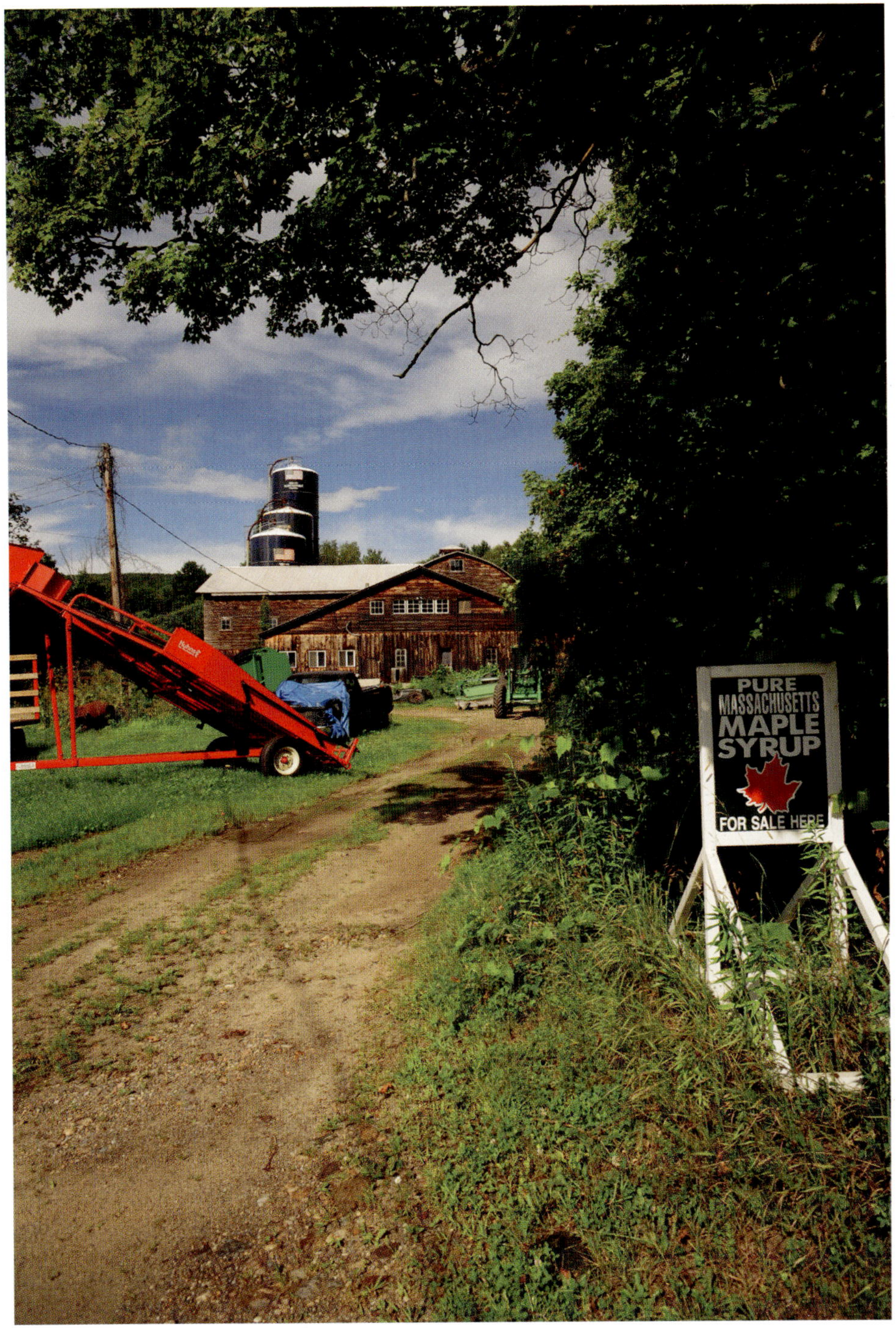

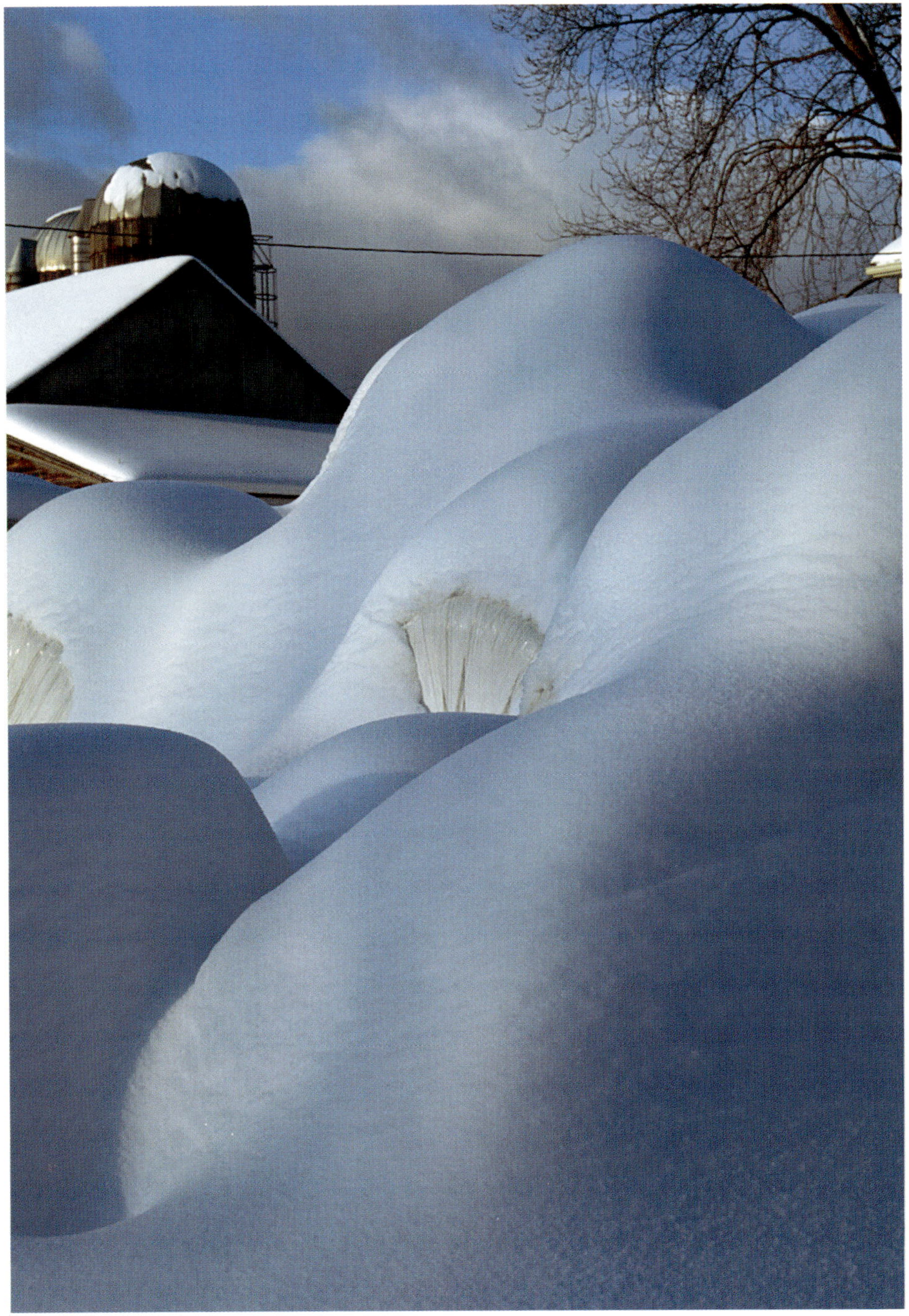

Covered hay bales lay under a thick fresh blanket of snow in Great Barrington.

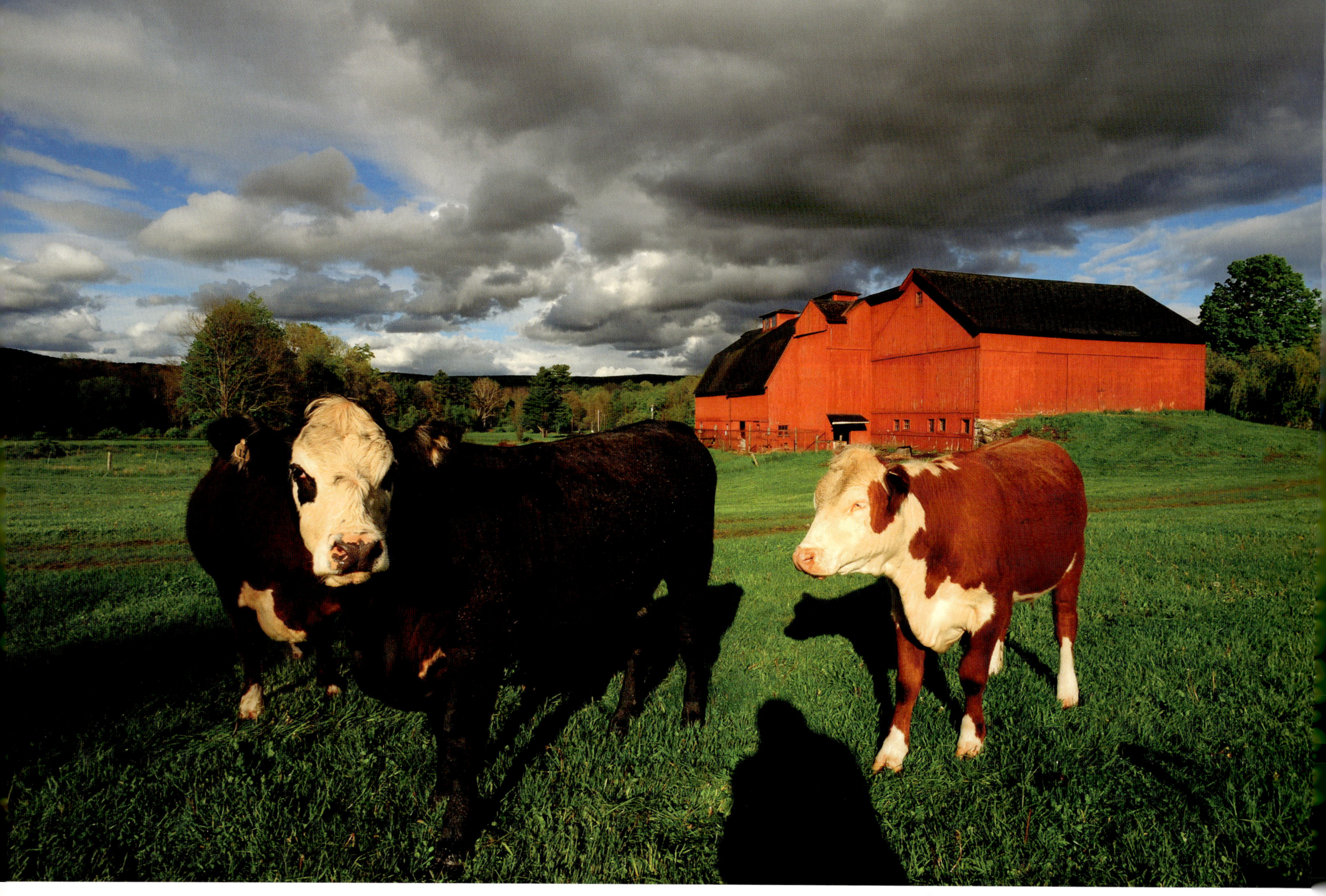

A last look at the barns on Division Street in Great Barrington, highlighted by late-day sun and the heavy clouds of a clearing storm. The younger gambrel-roof barn, which is set parallel to the road, and was not previously noted, was built between 1903 and 1905.

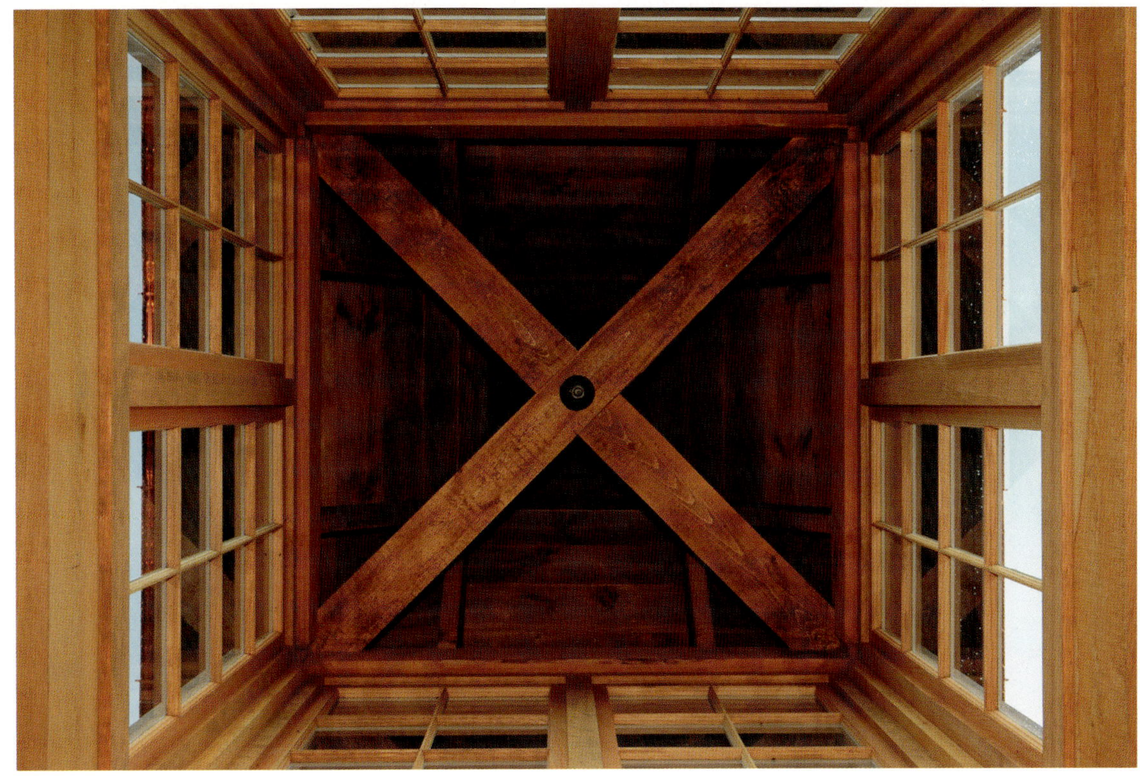

View up into the cupola that is a new feature in the restoration and relocation of this swing beam barn, brought to the Berkshires from Ontario by David Lanoue.

Bibliography

Hubka, Thomas C. "The Americanization of the Barn." *Blueprints,* Spring 1994. http://nbm.org (accessed April 2, 2008).

Leffingwell, Randy. *The American Barn*. Osceola, Wisconsin: Motorbooks International Publishers, 1997.

Majdalany, Ron. *The History of Seekonk Farm*. Self-Published, 2002

McLaughliin, Abraham. "Northeast's Battle Over Barns." *Christian Science Monitor,* February 13, 2003 http://www.csmonitor.com (accessed March 6, 2008)

Noble, Allen G., and Richard K. Cleek. *The Old Barn Book.* New Brunswick, New Jersey: Rutgers University Press, 1995.

Sloane, Eric. *An Age of Barns.* Stillwater, Minnesota: Voyageur Press, 2001.

Sobon, Jack. "The English Barn in America." *Timber Framing* 80 (June, 2006): 22-26.

Sobon, Jack, and Roger Schroeder, *Timber Frame Construction.* Pownal, Vermont: Storey Communications, 1984.

Visser, Thomas Durant. *Field Guide to New England Barns and Farm Buildings.* Hanover, New Hampshire: University Press of New England, 1997

Endnotes

1. Abraham McLaughlin, "Northeast's Battle Over Barns," *Christian Science Monitor* 13 February, 2003 <http:csmonitor.com/2003/0213/p03s01-ussc.htm>

2. *ibid.*

3. Eric Sloane, *An Age of Barns* (Stillwater, Minnesota: Voyageur Press, 2001) p.17

4. *ibid.* p.8

5. Jack A. Sobon, "The English Barn in America," *Timber Framing,* June, 2006, p.25

6. Thomas Durant Visser, *Field Guide to New England Farms and Barn Buildings* (Hanover, New Hampshire, University Press of New England, 1997) p. 30

7. *ibid* pp. 19-21

8. *ibid* pp. 24-30

9. *ibid* p. 37

10. *ibid* p. 39

11. *ibid* pp. 45-48

12. *ibid* p. 47

13. Thomas C. Hubka, "The Americanization of the Barn", *Blueprints Magazine* [Journal of the National Building Museum], Spring 1994, p.3

14. Allen G. Noble & Richard K. Cleek, *The Old Barn Book: A Field Guide to North American Barns & Other Structures* (New Brunswick, New Jersey: Rutgers university Press, 1995) pp. 22, 61, 78-79, 116-117

15. Sloane, *An Age of Barns*, p. 53

16. *ibid* p. 52

17. Viesser, *Field Guide to New England Farms and Barn Buildings*, pp. 125-138

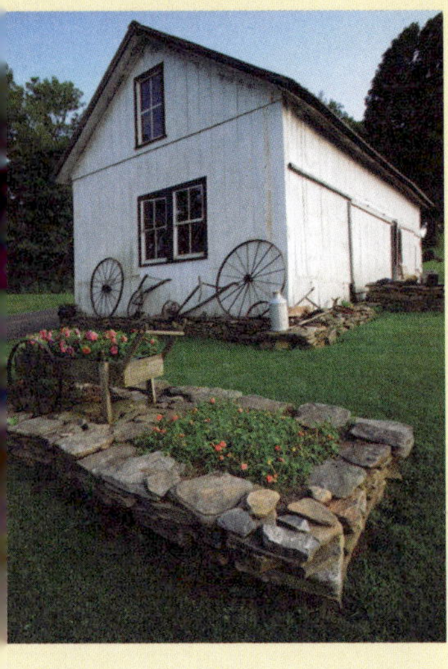
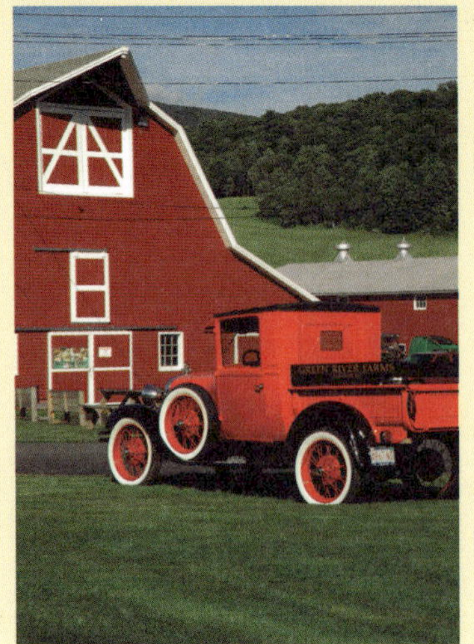
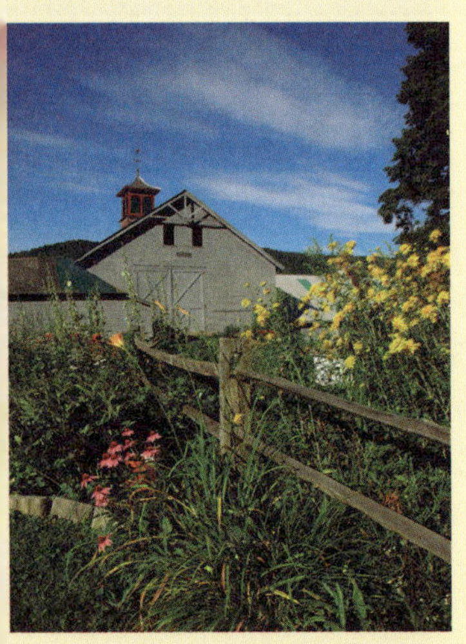
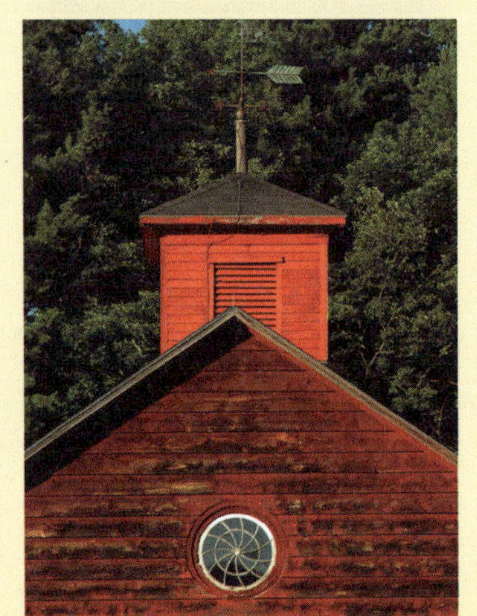
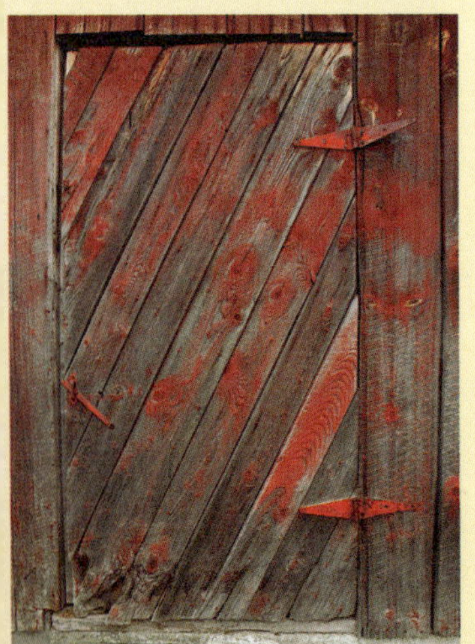
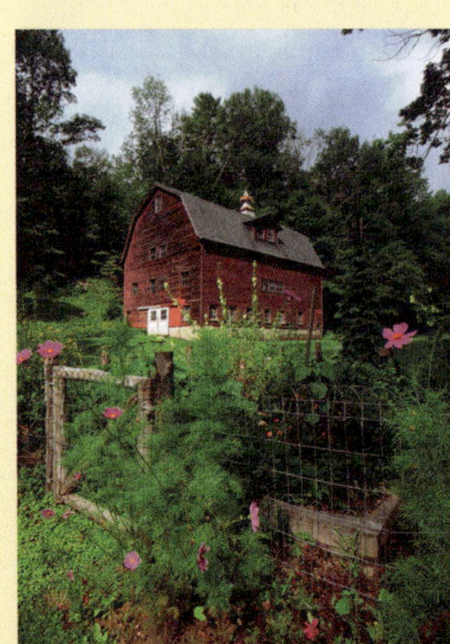

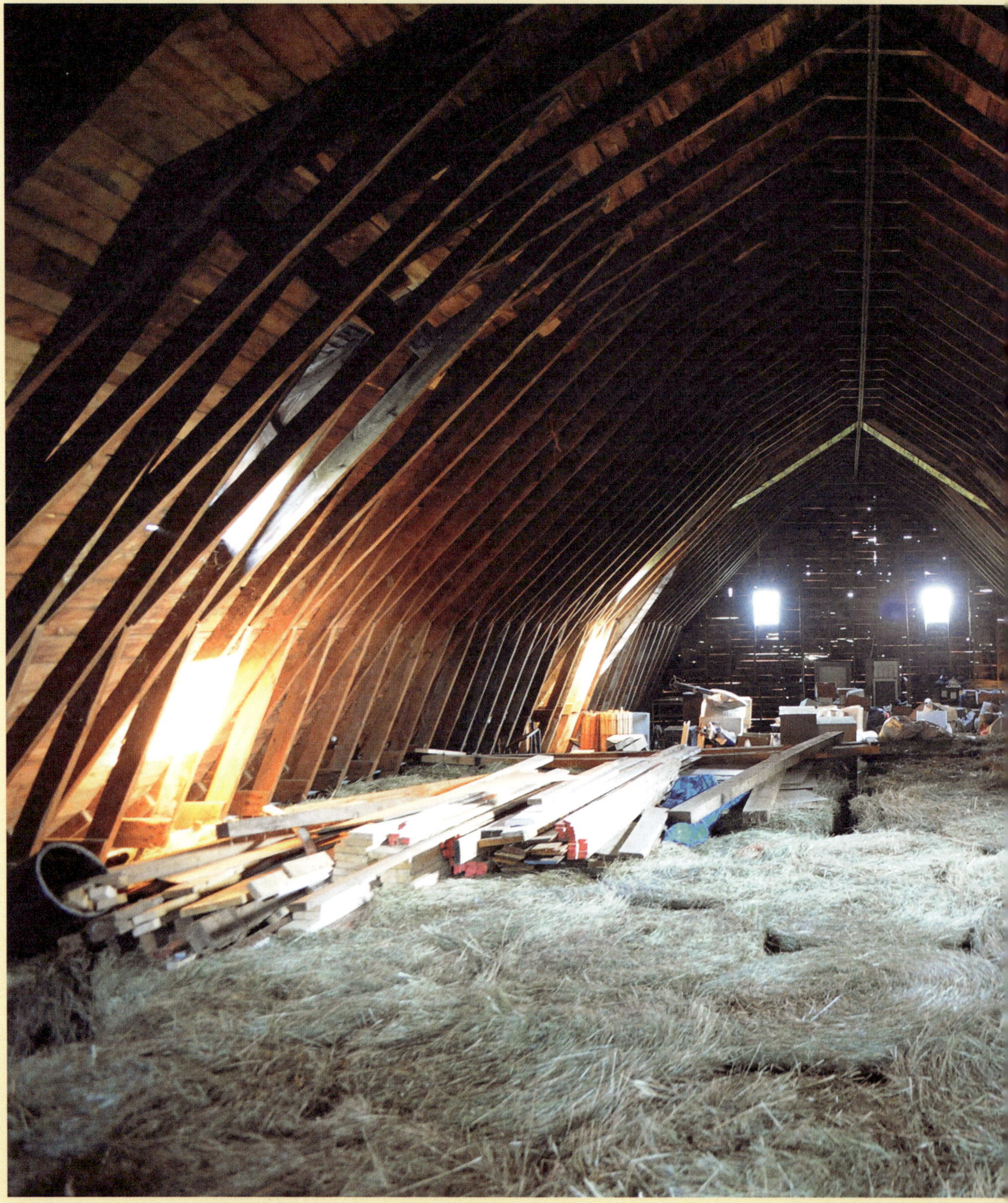